MEMOIRS
of the
American Mathematical Society

ISSN 0065-9266 (print)
ISSN 1947-6221 (online)

Number 1348

The Brunn-Minkowski Inequality and a Minkowski Problem for Nonlinear Capacity

Murat Akman
Jasun Gong
Jay Hineman
John Lewis
Andrew Vogel

January 2022 • Volume 275 • Number 1348 (second of 6 numbers)

AMERICAN MATHEMATICAL SOCIETY

Library of Congress Cataloging-in-Publication Data

Cataloging-in-Publication Data has been applied for by the AMS.
See http://www.loc.gov/publish/cip/.
DOI: https://doi.org/10.1090/memo/1348

Memoirs of the American Mathematical Society

This journal is devoted entirely to research in pure and applied mathematics.

Subscription information. Beginning with the January 2010 issue, *Memoirs* is accessible from `www.ams.org/journals`. The 2022 subscription begins with volume 275 and consists of six mailings, each containing one or more numbers. Subscription prices for 2022 are as follows: for paper delivery, US$1085 list, US$868 institutional member; for electronic delivery, US$955 list, US$764 institutional member. Upon request, subscribers to paper delivery of this journal are also entitled to receive electronic delivery. If ordering the paper version, add US$22 for delivery within the United States; US$85 for outside the United States. Subscription renewals are subject to late fees. See `www.ams.org/help-faq` for more journal subscription information. Each number may be ordered separately; *please specify number* when ordering an individual number.

Back number information. For back issues see `www.ams.org/backvols`.

Subscriptions and orders should be addressed to the American Mathematical Society, P. O. Box 845904, Boston, MA 02284-5904 USA. *All orders must be accompanied by payment.* Other correspondence should be addressed to 201 Charles Street, Providence, RI 02904-2213 USA.

Memoirs of the American Mathematical Society (ISSN 0065-9266 (print); 1947-6221 (online)) is published bimonthly (each volume consisting usually of more than one number) by the American Mathematical Society at 201 Charles Street, Providence, RI 02904-2213 USA. Periodicals postage paid at Providence, RI. Postmaster: Send address changes to Memoirs, American Mathematical Society, 201 Charles Street, Providence, RI 02904-2213 USA.

This publication is indexed in *Mathematical Reviews®*, *Zentralblatt MATH*, *Science Citation Index®*, *Science Citation Index™-Expanded*, *ISI Alerting Services℠*, *SciSearch®*, *Research Alert®*, *CompuMath Citation Index®*, *Current Contents®/Physical, Chemical & Earth Sciences*.
This publication is archived in *Portico* and *CLOCKSS*.
Printed in the United States of America.

♾ The paper used in this book is acid-free and falls within the guidelines established to ensure permanence and durability.
Visit the AMS home page at `https://www.ams.org/`

10 9 8 7 6 5 4 3 2 1 27 26 25 24 23 22 22

Contents

Abstract

In this article we study two classical potential-theoretic problems in convex geometry. The first problem is an inequality of Brunn-Minkowski type for a nonlinear capacity, $\mathrm{Cap}_{\mathcal{A}}$, where $\mathcal{A}$-capacity is associated with a nonlinear elliptic PDE whose structure is modeled on the p-Laplace equation and whose solutions in an open set are called $\mathcal{A}$-harmonic.

In the first part of this article, we prove the Brunn-Minkowski inequality for this capacity:

$$[\mathrm{Cap}_{\mathcal{A}}(\lambda E_1 + (1-\lambda)E_2)]^{\frac{1}{(n-p)}} \geq \lambda\, [\mathrm{Cap}_{\mathcal{A}}(E_1)]^{\frac{1}{(n-p)}} + (1-\lambda)\, [\mathrm{Cap}_{\mathcal{A}}(E_2)]^{\frac{1}{(n-p)}}$$

when $1 < p < n, 0 < \lambda < 1$, and E_1, E_2 are convex compact sets with positive $\mathcal{A}$-capacity. Moreover, if equality holds in the above inequality for some E_1 and E_2, then under certain regularity and structural assumptions on $\mathcal{A}$, we show that these two sets are homothetic.

In the second part of this article we study a Minkowski problem for a certain measure associated with a compact convex set E with nonempty interior and its $\mathcal{A}$-harmonic capacitary function in the complement of E. If μ_E denotes this measure, then the Minkowski problem we consider in this setting is that; for a given finite Borel measure μ on $\mathbb{S}^{n-1}$, find necessary and sufficient conditions for which there exists E as above with $\mu_E = \mu$. We show that necessary and sufficient conditions for existence under this setting are exactly the same conditions as in the classical Minkowski problem for volume as well as in the work of Jerison in [**J**] for

---

Received by the editor September 1, 2017, and, in revised form, October 8, 2018.

Article electronically published on December 13, 2021.

DOI: `https://doi.org/10.1090/memo/1348`

2020 *Mathematics Subject Classification.* Primary: 35J60,31B15,39B62,52A40,35J20,52A20,35J92.

Key words and phrases. The Brunn-Minkowski inequality, nonlinear capacities, inequalities and extremum problems, potentials and capacities, $\mathcal{A}$-harmonic PDEs, Minkowski problem, variational formula, Hadamard variational formula.

The first author is affiliated with the Department of Mathematical Sciences, University of Essex, Wivenhoe Park, Colchester, Essex CO4 3SQ, United Kingdom.

The second author is affiliated with the Mathematics Department, Fordham University, John Mulcahy Hall, Bronx, New York 10458-5165.

The third author is affiliated with Geometric Data Analysis, Durham, North Carolina 27707 AND the Department of Electrical and Computer Engineering, Duke University, Durham, North Carolina 27708.

The fourth author is affiliated with the Department of Mathematics, University of Kentucky, Lexington, Kentucky 40506.

The fifth author is affiliated with the Department of Mathematics, Syracuse University, Syracuse, New York 13244.

electrostatic capacity. Using the Brunn-Minkowski inequality result from the first part, we also show that this problem has a unique solution up to translation when $p \neq n-1$ and translation and dilation when $p = n-1$.

Part 1

The Brunn-Minkowski inequality for nonlinear capacity

CHAPTER 1

Introduction

The well-known Brunn-Minkowski inequality states that

$$(1.1)\qquad [\mathrm{Vol}(\lambda E_1+(1-\lambda)E_2)]^{\frac{1}{n}} \geq \lambda\,[\mathrm{Vol}(E_1)]^{\frac{1}{n}} + (1-\lambda)\,[\mathrm{Vol}(E_2)]^{\frac{1}{n}}$$

whenever E_1, E_2 are compact convex sets with nonempty interiors in $\mathbb{R}^n$ and $\lambda \in (0,1)$. Moreover, equality in (1.1) holds if and only if E_1 is a translation and dilation of E_2. For numerous applications of this inequality to problems in geometry and analysis see the classical book by Schneider [**Sc**] and the survey paper by Gardner [**G**]. Here $\mathrm{Vol}(\cdot)$ denotes the usual volume in $\mathbb{R}^n$ and the summation $(\lambda E_1 + (1-\lambda)E_2)$ should be understood as a vector sum(called *Minkowski addition*). (1.1) says that $[\mathrm{Vol}(\cdot)]^{1/n}$ is a concave function with respect to Minkowski addition. Inequalities of Brunn-Minkowski type have also been proved for other homogeneous functionals. For example, one can replace volume in (1.1) by "capacity" and in this case it was shown by Borell in [**B1**] that

$$(1.2)\qquad [\mathrm{Cap}_2(\lambda E_1+(1-\lambda)E_2)]^{\frac{1}{n-2}} \geq \lambda\,[\mathrm{Cap}_2(E_1)]^{\frac{1}{n-2}} + (1-\lambda)\,[\mathrm{Cap}_2(E_2)]^{\frac{1}{n-2}}$$

whenever E_1, E_2 are compact convex sets with nonempty interiors in $\mathbb{R}^n$, $n \geq 3$. Here Cap_2 denotes the *Newtonian capacity*. The exponents in this inequality and (1.1) differ as $\mathrm{Vol}(\cdot)$ is homogeneous of degree n whereas $\mathrm{Cap}_2(\cdot)$ is homogeneous of degree $n-2$. In [**B2**], Borell proved a Brunn-Minkowski type inequality for *logarithmic capacity*. The equality case in (1.2) was studied by Caffarelli, Jerison and Lieb in [**CJL**] and it was shown that equality in (1.2) holds if and only if E_2 is a translate and dilate of E_1 when $n \geq 3$. Jerison in [**J**] used that result to prove uniqueness in the *Minkowski problem* (see chapter 8 for the Minkowski problem). In [**CS**] Colesanti and Salani proved the p-capacitary version of (1.2) for $1 < p < n$. That is,

$$(1.3)\quad \left[\mathrm{Cap}_p(\lambda E_1+(1-\lambda)E_2)\right]^{\frac{1}{n-p}} \geq \lambda\left[\mathrm{Cap}_p(E_1)\right]^{\frac{1}{n-p}} + (1-\lambda)\left[\mathrm{Cap}_p(E_2)\right]^{\frac{1}{n-p}}$$

whenever E_1, E_2 are compact convex sets with nonempty interiors in $\mathbb{R}^n$, and $\mathrm{Cap}_p(\cdot)$ denotes the p-capacity of a set defined as

$$\mathrm{Cap}_p(E) = \inf\left\{ \int_{\mathbb{R}^n} |\nabla v|^p dx : \; v \in C_0^\infty(\mathbb{R}^n),\, v(x) \geq 1 \text{ for } x \in E \right\}.$$

It was also shown in the same paper that equality in (1.3) holds if and only if E_2 is a translate and dilate of E_1. In [**CC**], Colesanti and Cuoghi defined a logarithmic capacity for $p = n, n \geq 3$, and proved a Brunn-Minkowski type inequality for this capacity. In [**CNSXYZ**], a Minkowski problem was studied for p-capacity, $1 < p < 2$, using (1.3). See [**C**] for the torsional rigidity and first eigenvalue of the Laplacian versions of (1.1).

CHAPTER 2

Notation and statement of results

Let $n \geq 2$ and points in Euclidean n-space $\mathbb{R}^n$ be denoted by $y = (y_1, \dots, y_n)$. $\mathbb{S}^{n-1}$ will denote the unit sphere in $\mathbb{R}^n$. We write $e_m, 1 \leq m \leq n$, for the point in $\mathbb{R}^n$ with 1 in the m-th coordinate and 0 elsewhere. Let $\bar{E}, \partial E$, $\mathrm{diam}(E)$, be the closure, boundary, diameter, of the set $E \subset \mathbb{R}^n$ and we define $d(y, E)$ to be the distance from $y \in \mathbb{R}^n$ to E. Given two sets, $E, F \subset \mathbb{R}^n$ let

$$d_{\mathcal{H}}(E, F) = \max(\sup\{d(y, E) : y \in F\}, \sup\{d(y, F) : y \in E\})$$

be the Hausdorff distance between the sets $E, F \subset \mathbb{R}^n$. Also

$$E + F = \{x + y : x \in E, y \in F\}$$

is the Minkowski sum of E and F. We write $E + x$ for $E + \{x\}$ and set $\rho E = \{\rho y : y \in E\}$. Let $\langle \cdot, \cdot \rangle$ denote the standard inner product on $\mathbb{R}^n$ and let $|y| = \langle y, y \rangle^{1/2}$ be the Euclidean norm of y. Put

$$B(z, r) = \{y \in \mathbb{R}^n : |z - y| < r\} \quad \text{whenever } z \in \mathbb{R}^n,\ r > 0,$$

and dy denote Lebesgue n-measure on $\mathbb{R}^n$. Let $\mathcal{H}^k, 0 < k \leq n$, denote k-dimensional *Hausdorff measure* on $\mathbb{R}^n$ defined by

$$\mathcal{H}^k(E) = \lim_{\delta \to 0} \inf \left\{ \sum_j r_j^k;\ E \subset \cup B(x_j, r_j),\ r_j \leq \delta \right\}$$

where infimum is taken over all possible cover $\{B(x_j, r_j)\}_j$ of set E. If $O \subset \mathbb{R}^n$ is open and $1 \leq q \leq \infty$, then by $W^{1,q}(O)$ we denote the space of equivalence classes of functions h with distributional gradient $\nabla h = (h_{y_1}, \dots, h_{y_n})$, both of which are q-th power integrable on O. Let

$$\|h\|_{1,q} = \|h\|_q + \|\, |\nabla h|\, \|_q$$

be the norm in $W^{1,q}(O)$ where $\| \cdot \|_q$ is the usual Lebesgue q norm of functions in the Lebesgue space $L^q(O)$. Next let $C_0^\infty(O)$ be the set of infinitely differentiable functions with compact support in O and let $W_0^{1,q}(O)$ be the closure of $C_0^\infty(O)$ in the norm of $W^{1,q}(O)$. By $\nabla \cdot$ we denote the divergence operator.

DEFINITION 2.1. *Let $p, \alpha \in (1, \infty)$ and*

$$\mathcal{A} = (\mathcal{A}_1, \dots, \mathcal{A}_n) : \mathbb{R}^n \setminus \{0\} \to \mathbb{R}^n,$$

such that $\mathcal{A} = \mathcal{A}(\eta)$ has continuous partial derivatives in $\eta_k, 1 \leq k \leq n$, on $\mathbb{R}^n \setminus \{0\}$. We say that the function $\mathcal{A}$ belongs to the class $M_p(\alpha)$ if the following conditions *are satisfied whenever* $\xi \in \mathbb{R}^n$ *and* $\eta \in \mathbb{R}^n \setminus \{0\}$*:*

$$(i)\ \alpha^{-1}|\eta|^{p-2}|\xi|^2 \leq \sum_{i,j=1}^{n} \frac{\partial \mathcal{A}_i}{\partial \eta_j}(\eta)\xi_i\xi_j \ \text{and} \ \sum_{i,j=1}^{n} \left|\frac{\partial \mathcal{A}_i}{\partial \eta_j}\right| \leq \alpha|\eta|^{p-2},$$

$$(ii)\ \mathcal{A}(\eta) = |\eta|^{p-1}\mathcal{A}(\eta/|\eta|).$$

We put $\mathcal{A}(0) = 0$ and note that Definition 2.1 $(i), (ii)$ implies

$$\begin{aligned} c^{-1}(|\eta| + |\eta'|)^{p-2}\,|\eta - \eta'|^2 &\leq \langle \mathcal{A}(\eta) - \mathcal{A}(\eta'), \eta - \eta' \rangle \\ &\leq c|\eta - \eta'|^2(|\eta| + |\eta'|)^{p-2} \end{aligned} \tag{2.1}$$

whenever $\eta, \eta' \in \mathbb{R}^n \setminus \{0\}$.

DEFINITION 2.2. *Let* $p \in (1, \infty)$ *and let* $\mathcal{A} \in M_p(\alpha)$ *for some* α*. Given an open set* O *we say that* u *is* $\mathcal{A}$*-harmonic in* O *provided* $u \in W^{1,p}(G)$ *for each open* G *with* $\bar{G} \subset O$ *and*

$$\int \langle \mathcal{A}(\nabla u(y)), \nabla \theta(y) \rangle \, dy = 0 \quad \text{whenever } \theta \in W_0^{1,p}(G). \tag{2.2}$$

We say that u *is an* $\mathcal{A}$*-subsolution (*$\mathcal{A}$*-supersolution) in* O *if* $u \in W^{1,p}(G)$ *whenever* G *is as above and* (2.2) *holds with* $=$ *replaced by* $\leq$ $(\geq)$ *whenever* $\theta \in W_0^{1,p}(G)$ *with* $\theta \geq 0$*. As a short notation for* (2.2) *we write* $\nabla \cdot \mathcal{A}(\nabla u) = 0$ *in* O*.*

More about PDEs of this generalized type can be found in [**HKM**, Chapter 5] and [**A**, **ALV**, **LLN**, **LN4**]. If $\mathcal{A}(\eta) = |\eta|^{p-2}(\eta_1, \dots, \eta_n)$, and u is a weak solution relative to this $\mathcal{A}$ in O, then u is said to be p-harmonic in O.

REMARK 2.3. *We remark for* $O, \mathcal{A}, p, u,$ *as in Definition* 2.2 *that if* $F : \mathbb{R}^n \to \mathbb{R}^n$ *is the composition of a translation, and a dilation then*

$$\hat{u}(z) = u(F(z)) \text{ whenever } F(z) \in O \text{ is } \mathcal{A}\text{-harmonic in } F^{-1}(O).$$

Moreover, if $\tilde{F} : \mathbb{R}^n \to \mathbb{R}^n$ *is the composition of a translation, a dilation, and a rotation then*

$$\tilde{u}(z) = u(\tilde{F}(z)) \text{ is } \tilde{\mathcal{A}}\text{-harmonic in } \tilde{F}^{-1}(O) \text{ and } \tilde{\mathcal{A}} \in M_p(\alpha).$$

We shall use this remark numerous times in our proofs.

Let $E \subset \mathbb{R}^n$ be a compact convex set and let $\Omega = \mathbb{R}^n \setminus E$. Using (2.1), results in [**HKM**, Appendix 1], as well as Sobolev type limiting arguments, we show in Lemma 4.1 that if $\mathrm{Cap}_{\mathcal{A}}(E) > 0$, or equivalently $\mathcal{H}^{n-p}(E) = \infty$, then there exists a unique continuous function $u \not\equiv 1, 0 < u \leq 1$, on $\mathbb{R}^n$ satisfying

$$\begin{aligned} &(a)\ u \text{ is } \mathcal{A}\text{-harmonic in } \Omega, \\ &(b)\ u \equiv 1 \text{ on } E, \\ &(c)\ |\nabla u| \in L^p(\mathbb{R}^n) \text{ and } u \in L^{p^*}(\mathbb{R}^n) \text{ for } p^* = \frac{np}{n-p}. \end{aligned} \tag{2.3}$$

We put

$$\mathrm{Cap}_{\mathcal{A}}(E) = \int_{\Omega} \langle \mathcal{A}(\nabla u), \nabla u \rangle \, dy$$

and call $\mathrm{Cap}_{\mathcal{A}}(E)$, the $\mathcal{A}$*-capacity of* E while u is the $\mathcal{A}$-capacitary function corresponding to E in Ω. We note that this definition is a slight extension of the usual definition of *capacity*. However in case,

$$\mathcal{A}(\eta) = p^{-1}\nabla f(\eta) \quad \text{on } \mathbb{R}^n \setminus \{0\}$$

then from $p-1$ homogeneity in Definition 2.1 (ii) it follows that

$$f(t\eta) = t^p f(\eta) \quad \text{whenever } t > 0 \text{ and } \eta \in \mathbb{R}^n \setminus \{0\}.$$

In this case, using Euler's formula, one gets the usual definition of capacity relative to f. That is,

$$\mathrm{Cap}_{\mathcal{A}}(E) = \inf \left\{ \int_{\mathbb{R}^n} f(\nabla\psi(y))dy \, : \psi \in C_0^\infty(\mathbb{R}^n) \text{ with } \psi \geq 1 \text{ on } E \right\}.$$

See chapter 5 in [**HKM**] for more about this definition of capacity in terms of such f. In case

$$\mathcal{A}(\eta) = |\eta|^{p-2}(\eta_1, \ldots, \eta_n)$$

so we have $f(\eta) = p^{-1}|\eta|^p$ then the above capacity will be denoted by $\mathrm{Cap}_p(E)$ and called the p-capacity of E. Note from (2.1) with $\eta' = 0$, that

$$c^{-1}\,\mathrm{Cap}_p(E) \leq \mathrm{Cap}_{\mathcal{A}}(E) \leq c\,\mathrm{Cap}_p(E) \tag{2.4}$$

where c depends only on α, p, and n. From Remark 2.3 and uniqueness of u in (2.3), we observe for $z \in \mathbb{R}^n$ and $\rho > 0$, that if $\tilde{E} = \rho E + z$, then

(2.5)

(a') $\mathrm{Cap}_{\mathcal{A}}(\rho E + z) = \rho^{n-p}\,\mathrm{Cap}_{\mathcal{A}}(E)$,

(b') $\tilde{u}(x) = u((x-z)/\rho)$, for $x \in \mathbb{R}^n \setminus \tilde{E}$, is the $\mathcal{A}$-capacitary function for $\tilde{E}$.

Observe from (2.5) (a') that for $z \in \mathbb{R}^n$ and $R > 0$,

$$\mathrm{Cap}_{\mathcal{A}}(B(z,R)) = c_1 R^{n-p} \tag{2.6}$$

where c_1 depends only on p, n, α.

In the first part of this article, we prove the following Brunn-Minkowski type theorem for $\mathcal{A}$-capacities:

THEOREM A. *Let E_1, E_2 be compact convex sets in $\mathbb{R}^n$ satisfying $\mathrm{Cap}_{\mathcal{A}}(E_i) > 0$ for $i = 1, 2$. If $1 < p < n$ is fixed, $\mathcal{A}$ is as in Definition 2.1, and $\lambda \in [0,1]$, then*

(2.7)

$$[\mathrm{Cap}_{\mathcal{A}}(\lambda E_1 + (1-\lambda)E_2)]^{\frac{1}{(n-p)}} \geq \lambda\,[\mathrm{Cap}_{\mathcal{A}}(E_1)]^{\frac{1}{(n-p)}} + (1-\lambda)\,[\mathrm{Cap}_{\mathcal{A}}(E_2)]^{\frac{1}{(n-p)}}.$$

If equality holds in (2.7) and

(2.8)

(i) *There exists $1 \leq \Lambda < \infty$ such that* $\left| \dfrac{\partial \mathcal{A}_i}{\partial \eta_j}(\eta) - \dfrac{\partial \mathcal{A}_i}{\partial \eta'_j}(\eta') \right| \leq \Lambda\,|\eta - \eta'||\eta|^{p-3}$

whenever $0 < \frac{1}{2}|\eta| \leq |\eta'| \leq 2|\eta|$ *and* $1 \leq i, j \leq n$,

(ii) $\mathcal{A}_i(\eta) = \dfrac{\partial f}{\partial \eta_i}$ *for* $1 \leq i \leq n$ *where* $f(t\eta) = t^p f(\eta)$ *when* $t > 0$ *and* $\eta \in \mathbb{R}^n \setminus \{0\}$

then E_2 is a translation and dilation of E_1.

To briefly outline the proof of Theorem A, in chapter 3 we list some basic properties of $\mathcal{A}$-harmonic functions which will be used in the proof of Theorem A. We then use these properties in chapters 4 and 5 to prove inequality (2.7). The last sentence in Theorem A regarding the case of equality in (2.7) is proved in chapter 6.

As for the main steps in our proof, after the preliminary material, we show in Lemma 4.4 that if u is a nontrivial $\mathcal{A}$-harmonic capacitary function for a compact convex set E, then $\{x : u(x) > t\}$ is convex whenever $0 < t < 1$. The proof uses a maximum principle type argument of Gabriel in [**Ga**] to show that if u does not have levels bounding a convex domain, then a certain function has an absolute maximum in $\mathbb{R}^n \setminus E$, from which one obtains a contradiction. This argument was later used by the fourth named author of this article in [**L**] in the p-Laplace setting and also a variant of it was used by Borell in [**B1**] (see [**BLS**] for recent applications). After proving Lemma 4.1 we use an analogous argument to prove (2.7). Our proof of equality in Theorem A is inspired by the proof in [**CS**] which in turn uses some ideas of Longinetti in [**Lo**]. In particular, Lemma 2 in [**CS**] plays an important role in our proof. Unlike these authors though, we do not convert the PDE for u_1, u_2 into one for the support functions of their levels, essentially because our PDE is not rotationally invariant. The arguments we use require a priori knowledge that the levels of u_1, u_2 have positive curvatures. We can show this near ∞ when $\mathcal{A} = \nabla f$, as in Theorem A, by comparing u_1, u_2 with their respective "fundamental solutions" (see Lemma 6.1) which can be calculated more or less directly. A unique continuation argument then gives Theorem A. This argument does not work for a general $\mathcal{A}$. In this case a method first used by Korevaar in [**K**] and after that by various authors (see [**BGMX**]) appears promising, although rather tedious and at the expense of assuming more regularity on $\mathcal{A}$ for handling the case of equality in Theorem A. Finally, we mention that our main purpose in working on the Brunn-Minkowski inequality is to prepare a background for our investigation of a Minkowski problem when $\mathcal{A} = \nabla f$ and $1 < p < n$ (see Theorem B in chapter 8).

CHAPTER 3

Basic estimates for $\mathcal{A}$-harmonic functions

In this chapter we state some fundamental estimates for $\mathcal{A}$-harmonic functions. Concerning constants, unless otherwise stated, in this chapter, and throughout the paper, c will denote a positive constant ≥ 1, not necessarily the same at each occurrence, depending at most on p, n, α, Λ which sometimes we refer to as depending on the data. In general, $c(a_1, \dots, a_m)$ denotes a positive constant ≥ 1, which may depend at most on the data and $a_1, \dots, a_m$, not necessarily the same at each occurrence. If $B \approx C$ then B/C is bounded from above and below by constants which, unless otherwise stated, depend at most on the data. Moreover, we let $\max_F \tilde{u}$, $\min_F \tilde{u}$ be the essential supremum and infimum of $\tilde{u}$ on F whenever $F \subset \mathbb{R}^n$ and whenever $\tilde{u}$ is defined on F.

LEMMA 3.1. *Given $p, 1 < p < n$, assume that $\tilde{\mathcal{A}} \in M_p(\alpha)$ for some $\alpha > 1$. Let $\tilde{u}$ be a positive $\tilde{\mathcal{A}}$-harmonic function in $B(w, 4r)$, $r > 0$. Then*

$$(3.1)\qquad \begin{aligned} &(i)\ r^{p-n} \int_{B(w,r/2)} |\nabla \tilde{u}|^p \, dy \,\leq\, c\,(\max_{B(w,r)} \tilde{u})^p, \\ &(ii)\ \max_{B(w,r)} \tilde{u} \leq c \min_{B(w,r)} \tilde{u}. \end{aligned}$$

Furthermore, there exists $\tilde{\sigma} = \tilde{\sigma}(p, n, \alpha) \in (0, 1)$ such that if $x, y \in B(w, r)$, then

$$(iii)\ \ |\tilde{u}(x) - \tilde{u}(y)| \leq c \left(\frac{|x-y|}{r} \right)^{\tilde{\sigma}} \max_{B(w,2r)} \tilde{u}.$$

PROOF. A proof of this lemma can be found in [**S**]. □

LEMMA 3.2. *Let $p, n, \tilde{\mathcal{A}}, \alpha, w, r, \tilde{u}$ be as in Lemma 3.1. Then $\tilde{u}$ has a representative locally in $W^{1,p}(B(w, 4r))$, with Hölder continuous partial derivatives in $B(w, 4r)$ (also denoted $\tilde{u}$), and there exists $\tilde{\beta} \in (0, 1], c \geq 1$, depending only on p, n, α, such that if $x, y \in B(w, r)$, then*

(3.2)

$$(\hat{a})\ c^{-1}\,|\nabla \tilde{u}(x) - \nabla \tilde{u}(y)| \,\leq\, (|x-y|/r)^{\tilde{\beta}} \max_{B(w,r)} |\nabla \tilde{u}| \leq c\,r^{-1}\,(|x-y|/r)^{\tilde{\beta}}\,\tilde{u}(w).$$

$$(\hat{b})\ \int_{B(w,r)} \sum_{i,j=1}^{n} |\nabla \tilde{u}|^{p-2}\,|\tilde{u}_{x_i x_j}|^2 dy \leq c r^{(n-p-2)} \tilde{u}(w).$$

If

$$\gamma\, r^{-1} \tilde{u} \leq |\nabla \tilde{u}| \leq \gamma^{-1} r^{-1} \tilde{u} \quad on \quad B(w, 2r)$$

for some $\gamma \in (0, 1)$ and (2.8) (i) holds then $\tilde{u}$ has Hölder continuous second partial derivatives in $B(w, r)$ and there exists $\tilde{\theta} \in (0, 1), \bar{c} \geq 1$, depending only on the data *and* γ *such that*

(3.3)

$$\left[\sum_{i,j=1}^{n}(\tilde u_{x_ix_j}(x)-\tilde u_{y_iy_j}(y))^2\right]^{1/2}\leq \bar c(|x-y|/r)^{\tilde\theta}\max_{B(w,r)}\left(\sum_{i,j=1}^{n}|\tilde u_{x_ix_j}|\right)$$

$$\leq \bar c^2 r^{-n/2}(|x-y|/r)^{\tilde\theta}\left(\sum_{i,j=1}^{n}\int_{B(w,2r)}\tilde u_{x_ix_j}^2\,dx\right)^{1/2}$$

$$\leq \bar c^3 r^{-2}(|x-y|/r)^{\tilde\theta}\tilde u(w).$$

whenever $x,y\in B(w,r/2)$.

PROOF. A proof of (3.2) can be found in [**T**]. Also, (3.3) follows from (3.2), the added assumptions, and Schauder type estimates (see [**GT**]). □

LEMMA 3.3. *Fix* $p, 1<p<n$, *assume that* $\tilde{\mathcal{A}}\in M_p(\alpha)$, *and let* $\tilde E\subset B(0,R)$, *for some* $R>0$, *be a compact convex set with* $\mathrm{Cap}_{\mathcal{A}}(\tilde E)>0$. *Let* $\zeta\in C_0^\infty(B(0,2R))$ *with* $\zeta\equiv 1$ *on* $B(0,R)$. *If* $0\leq\tilde u$ *is* $\tilde{\mathcal{A}}$-*harmonic in* $B(0,4R)\setminus\tilde E$, *and* $\tilde u\zeta\in W_0^{1,p}(B(0,4R)\setminus\tilde E)$, *then* $\tilde u$ *has a continuous extension to* $B(0,4R)$ *obtained by putting* $\tilde u\equiv 0$ *on* $\tilde E$. *Moreover, if* $0<r<R$ *and* $w\in\partial\tilde E$ *then*

$$(i)\ r^{p-n}\int_{B(w,r)}|\nabla\tilde u|^p\,dy\leq c\left(\max_{B(w,2r)}\tilde u\right)^p. \tag{3.4}$$

Furthermore, there exists $\hat\sigma=\hat\sigma(p,n,\alpha,\tilde E)\in(0,1)$ *such that if* $x,y\in B(w,r)$ *and* $0<r<\mathrm{diam}(\tilde E)$ *then*

$$(ii)\ |\tilde u(x)-\tilde u(y)|\leq c\left(\frac{|x-y|}{r}\right)^{\hat\sigma}\max_{B(w,2r)}\tilde u.$$

PROOF. Here (i) is a standard Caccioppoli inequality. To prove (ii) we note that necessarily $\mathcal{H}^{n-p}(\tilde E)=\infty$, as follows from (2.4) and Theorem 2.27 in [**HKM**]. From this note, as well as convexity and compactness of $\tilde E$, we deduce that

$$\mathcal{H}^l(B(y,r)\cap\tilde E)\approx r^l$$

for some positive integer $l>n-p$, whenever $y\in\tilde E$ and $0<r<\mathrm{diam}(\tilde E)$. Constants depend on $\tilde E$ but are independent of r,z. Using this fact and metric properties of certain capacities in chapter 2 of [**HKM**], it follows that

$$\mathrm{Cap}_p(B(y,r)\cap\tilde E)\approx r^{n-p}\quad\text{whenever } 0<r<\mathrm{diam}(\tilde E)\text{ and } y\in\tilde E \tag{3.5}$$

where constants depend on α,p,n and $\tilde E$. Now (ii) for $y\in\tilde E$ follows from (3.5) and essentially Theorem 6.18 in [**HKM**]. Combining this fact with (3.1) (iii) we now obtain (ii). □

LEMMA 3.4. *Let* $\tilde{\mathcal{A}},p,n,\tilde E,R,\tilde u$ *be as in Lemma 3.3. Then there exists a unique finite positive Borel measure* $\tilde\mu$ *on* $\mathbb{R}^n$, *with support contained in* $\tilde E$ *such that if* $\phi\in C_0^\infty(B(0,2R))$ *then*

$$(i)\ \int\langle\tilde{\mathcal{A}}(\nabla\tilde u(y)),\nabla\phi(y)\rangle\,dy=-\int\phi\,d\tilde\mu. \tag{3.6}$$

Moreover, if $0 < r \leq R$ *and* $w \in \partial \tilde{E}$ *then there exists* $c \geq 1$, *depending only on the data such that*

$$(ii)\ r^{p-n} \tilde{\mu}(B(w,r)) \leq c \max_{B(w,2r)} \tilde{u}^{p-1}.$$

PROOF. For the proof of (i) see Theorem 21.2 in [**HKM**]. (ii) follows from (2.1) with $\eta' = (0, \dots, 0)$, Hölder's inequality, and (3.4) (i) using a test function, ϕ, with $\phi \equiv 1$ on $\tilde{E}$. □

CHAPTER 4

Preliminary reductions for the proof of Theorem A

Throughout this chapter we assume that E is a compact convex set with $0 \in E$, $\text{diam}(E) = 1$, and $\text{Cap}_{\mathcal{A}}(E) > 0$. We begin with

LEMMA 4.1. *For fixed $p, 1 < p < n$, there exists a unique locally Hölder continuous u on $\mathbb{R}^n$ satisfying* (2.3).

PROOF. Given a positive integer $m \geq 4$, let u_m be the $\mathcal{A}$-harmonic function in $B(0,m) \setminus E$ with u_m in $W_0^{1,p}(B(0,m))$ and $u_m = 1$ on E in the $W^{1,p}$ Sobolev sense. Existence of u_m is proved in [**HKM**, Corollary 17.3, Appendix 1]. From Lemma 3.3 with

$$\tilde{\mathcal{A}}(\eta) = -\mathcal{A}(-\eta) \quad \text{whenever } \eta \in \mathbb{R}^n,$$

and $\tilde{u} = 1 - u_m$ we see that u_m has a Hölder continuous extension to $B(0,m)$ with $u_m = 1$ on E. From Sobolev's theorem, (2.1), and results for certain p type capacities from [**HKM**] we see there exists $c = c(p,n)$ such that if $p^* = \frac{np}{n-p}$ then

$$\|u_m\|_{p^*} \leq c \, \| |\nabla u_m| \|_p \leq c^2 \tag{4.1}$$

where the norms are relative to $L^q(\mathbb{R}^n), q \in \{p^*, p\}$. From Lemmas 3.1, 3.3 we see that u_m is locally Hölder continuous on compact subsets of $B(0,m)$ with exponent and constant that is independent of m while ∇u_m is locally Hölder continuous on compact subsets of $B(0,m) \setminus E$ again with exponent and constant that is independent of m. Using these facts and Ascoli's theorem we see there exists a subsequence $\{u_{m_k}\}$ of $\{u_m\}$ with

$$\begin{aligned} &\{u_{m_k}, \nabla u_{m_k}\} \text{ converging to } \{u, \nabla u\} \text{ as } m_k \to \infty \\ &\qquad\qquad \text{uniformly on compact subsets of } \mathbb{R}^n, \mathbb{R}^n \setminus E, \text{ respectively.} \end{aligned}$$

From this fact, the Fatou's lemma, and Definition 2.2 we see that u is continuous on $\mathbb{R}^n$, $\mathcal{A}$-harmonic in $\mathbb{R}^n \setminus E$ with $u \equiv 1$ on E, and (4.1) holds with u_m replaced by u. Thus u satisfies (2.3).

To prove uniqueness we note from Harnack's inequality in (3.1) (ii) that for $|x| \geq 2$,

$$u(x)^{p^*} \leq \tilde{c} \, |x|^{-n} \int_{\mathbb{R}^n \setminus B(x,|x|/2)} u^{p^*} dy \leq \tilde{c}^2 \, |x|^{-n} \tag{4.2}$$

where $\tilde{c}$ has the same dependence as the constant in (4.1). If v also satisfies (2.3), then (4.2) holds with u replaced by v so from the usual Sobolev type limiting arguments we see for each $\epsilon > 0$ that $\theta = \max(|u - v| - \epsilon, 0)$ can be used as a test function in (2.2) for u and v. Doing this and using (2.1), it follows for some $c \geq 1$, depending only on the data that

(4.3)
$$\int_{\{|u-v|>\epsilon\}} (|\nabla u| + |\nabla v|)^{p-2} |\nabla u - \nabla v|^2 dy \leq c \int_{\mathbb{R}^n \setminus E} \langle \mathcal{A}(\nabla u) - \mathcal{A}(\nabla v), \nabla \theta \rangle \, dy$$
$$= 0.$$

Letting $\epsilon \to 0$ we conclude first from (4.3) that $u - v$ is constant on Ω and then from (2.3) (b) that $u \equiv v$. $\square$

Throughout the rest of this chapter, we assume u is the $\mathcal{A}$-capacitary function for E and a fixed $p, 1 < p < n$. Let μ be the measure associated with $\tilde{u} = 1 - u$, where $\tilde{u}$ is $\tilde{\mathcal{A}}(\eta) = -\mathcal{A}(-\eta)$-harmonic, as in Lemma 3.4. Next we prove

LEMMA 4.2. *For $\mathcal{H}^1$ almost every $t \in (0,1)$*

(4.4)
$$(a)\ \mu(E) = \mathrm{Cap}_{\mathcal{A}}(E) = \int_{\{u=t\}} \langle \mathcal{A}(\nabla u(y)), \nabla u(y)/|\nabla u(y)| \rangle \, d\mathcal{H}^{n-1}$$
$$= t^{-1} \int_{\{u<t\}} \langle \mathcal{A}(\nabla u(y)), \nabla u(y) \rangle \, dy.$$

(b) There exists $c \geq 1$, depending only on p, n, α, so that
$$c^{-1} \mathrm{Cap}_{\mathcal{A}}(E) |x|^{(p-n)} \leq u(x)^{p-1} \leq c |x|^{(p-n)} \mathrm{Cap}_{\mathcal{A}}(E) \quad \text{whenever } |x| \geq 2.$$

PROOF. To prove (4.4) (a) fix $R \geq 4$ and let $0 \leq \psi \in C_0^\infty(B(0,2R))$ with $\psi \equiv 1$ on $B(0,R)$ and $|\nabla \psi| \leq c/R$. Using Sobolev type estimates we see that $\phi = u\psi$ can be used as a test function in (3.6) (i) with $\tilde{u}, \tilde{\mathcal{A}}$ as above. We get

$$\begin{aligned}
\mu(E) &= \int_{\mathbb{R}^n} \langle \mathcal{A}(\nabla u), \nabla(u\psi) \rangle dy \\
&= \int_{\mathbb{R}^n} \langle \mathcal{A}(\nabla u), \nabla u \rangle \psi dy + \int_{\mathbb{R}^n} u \langle \mathcal{A}(\nabla u), \nabla \psi \rangle dy = T_1 + T_2.
\end{aligned} \tag{4.5}$$

From the definition of $\mathcal{A}$-capacity we have

$$T_1 \to \mathrm{Cap}_{\mathcal{A}}(E) \quad \text{as } R \to \infty. \tag{4.6}$$

Also, from (2.1) with $\eta' = (0, \ldots, 0)$ and Hölder's inequality, we deduce that

$$\begin{aligned}
|T_2| &\leq cR^{-1} \int_{B(0,2R) \setminus B(0,R)} u |\nabla u|^{p-1} dy \\
&\leq c^2 R^{-p} \int_{B(0,2R) \setminus B(0,R)} u^p dy + c^2 \int_{B(0,2R) \setminus B(0,R)} |\nabla u|^p dy.
\end{aligned} \tag{4.7}$$

Clearly, the last integral on the far right $\to 0$ as $R \to \infty$, thanks to (2.3) (c). Moreover, from Hölder's inequality and (4.1) for u, we have for some $\hat{c} = \hat{c}(p,n)$,

$$R^{-p} \int_{B(0,2R) \setminus B(0,R)} u^p dy \leq \hat{c} \left[\int_{B(0,2R) \setminus B(0,R)} u^{p^*} dy \right]^{p/p^*} \to 0 \text{ as } R \to \infty. \tag{4.8}$$

Using (4.8) in (4.7) we see first that $T_2 \to 0$ as $R \to \infty$ and then from (4.5), (4.6) that $\mu(E) = \mathrm{Cap}_{\mathcal{A}}(E)$. Next, given $\epsilon > 0$ and $t > 4\epsilon$, let $k \geq 0$ be infinitely differentiable on $\mathbb{R}$ with

$$k(x) = \begin{cases} 1 & \text{when } x \in [t+\epsilon, \infty), \\ 0 & \text{when } x \in (-\infty, t-\epsilon]. \end{cases}$$

Then using $k \circ u$ as a test function in (3.6) (i) we find that

$$\begin{aligned} \mu(E) &= \int_{\mathbb{R}^n} \langle \mathcal{A}(\nabla u), \nabla u \rangle (k' \circ u) dy \\ &= \int_{t-\epsilon}^{t+\epsilon} \left(\int_{\{u=s\}\cap\{|\nabla u|>0\}} \langle \mathcal{A}(\nabla u), \nabla u/|\nabla u| \rangle d\mathcal{H}^{n-1} \right) k'(s) ds \end{aligned} \tag{4.9}$$

where we have used the coarea theorem (see [**EG**, Section 3, Theorem 1]) to get the last integral. Let

$$I(s) = \int_{\{u=s\}\cap\{|\nabla u|>0\}} \langle \mathcal{A}(\nabla u), \nabla u/|\nabla u| \rangle d\mathcal{H}^{n-1}.$$

Then (4.9) can be written as

$$\mu(E) = I(t) + \int_{t-\epsilon}^{t+\epsilon} [I(s) - I(t)] k'(s) ds. \tag{4.10}$$

From (2.1), (2.3) (c), and the coarea theorem once again we see that I is integrable on $(0,1)$. Using this fact and the Lebesgue differentiation theorem we find that the integral in (4.10) $\to 0$ as $\epsilon \to 0$ for almost every $t \in (0,1)$. It remains to prove the final inequality in (4.4) (a). To accomplish this, replace t by τ in the far-right boundary integral in (4.4) (a), integrate from 0 to t and use the coarea theorem once again.

To prove (4.4) (b) we note that if $a_1 \le u \le b_1$ on $\partial B(0,\rho)$ for $\rho \ge 4$, then from (2.6) and (2.4) we deduce that

$$\begin{aligned} c\rho^{n-p} = \operatorname{Cap}_p(B(0,\rho)) &\le a_1^{-p} \int_{\mathbb{R}^n \setminus B(0,\rho)} |\nabla u|^p dy \\ &\le c' a_1^{-p} \int_{\{u \le b_1\}} \langle \mathcal{A}(\nabla u), \nabla u \rangle dy. \end{aligned} \tag{4.11}$$

Using (4.4) (a), (4.11), and Harnack's inequality we see for almost every a_1, b_1 with

$$\min_{\partial B(0,\rho)} u \le 2a_1 \quad \text{and} \quad \max_{\partial B(0,\rho)} u \ge b_1/2$$

we have

$$\rho^{n-p} \le c_- \, a_1^{-p} b_1 \operatorname{Cap}_{\mathcal{A}}(E) \le c_-^2 \, b_1^{1-p} \operatorname{Cap}_{\mathcal{A}}(E)$$

where c_- depends only on p, n, α. This inequality implies the right-hand inequality in (4.4) (b). To get the left-hand inequality in (4.4) (b) for given $x, |x| \ge 4$, let ψ be as in (4.5) with $R = |x|$. Using ψ as a test function in (3.6) (i) and using (2.1),

Hölder's inequality, Lemma 3.1 (i), and Harnack's inequality we obtain

$$\begin{aligned}|x|^{p-n}\operatorname{Cap}_{\mathcal{A}}(E) = |x|^{p-n}\mu(E) &\leq c\,|x|^{p-n-1}\int_{\{|x|<y<2|x|\}}|\nabla u|^{p-1}dy\\ &\leq c^2|x|^{(p-n)(1-1/p)}\left(\int_{\{|x|<y<2|x|\}}|\nabla u|^p dy\right)^{1-1/p}\\ &\leq c^3\left(\max_{\{|x|/2<|y|<4|x|\}} u\right)^{p-1} \leq c^4 u(x)^{p-1}\end{aligned}$$

which yields the left-hand inequality in (4.4) (b). The proof of Lemma 4.2 is now complete. □

For the following lemmas, let $\Omega = \mathbb{R}^n \setminus E$.

LEMMA 4.3. *If there exists $r_0 > 0$ and $z \in E$ with $B(z, r_0) \subset E$ then there is $c_* \geq 1$, depending only on p, n, α, r_0 such that*

$$\text{(4.12)}\qquad \begin{aligned}&(a) \qquad c_*\langle\nabla u(x), z-x\rangle \geq u(x) \quad \textit{whenever } x \in \Omega,\\ &(b) \qquad c_*^{-1}\,|x|^{\frac{1-n}{p-1}} \leq |\nabla u(x)| \leq c_*\,|x|^{\frac{1-n}{p-1}} \quad \textit{whenever } |x| \geq 4.\end{aligned}$$

PROOF. We may assume that $z = 0$ thanks to Remark 2.3 and the fact that (4.12) is invariant under translation. Let

$$v(x) = \frac{u(x) - u(\lambda x)}{\lambda - 1} - \frac{u(x)}{\breve{c}}$$

when $x \in \bar{\Omega}$ and $1 < \lambda \leq 11/10$. We claim that if $\breve{c} = \breve{c}(p, n, \alpha, r_0) \geq 1$ is large enough, then

$$\text{(4.13)}\qquad v(x) \geq 0 \quad \text{whenever} \quad x \in \bar{\Omega}.$$

From the maximum principle for $\mathcal{A}$-harmonic functions, we see that it suffices to prove (4.13) when $x \in \partial E$ or equivalently that

$$\text{(4.14)}\qquad 1 - u(\lambda x) \geq c^{-1}(\lambda - 1)|x| \quad \text{whenever} \quad x \in \partial E$$

for some $c = c(p, n, \alpha, r_0) \geq 1$ since $r_0 \leq |x| \leq 1$ and $x \in \partial E$. We also note that

$$d(\lambda x, E) \leq (\lambda - 1)|x| \leq c' d(\lambda x, \partial E) \quad \text{whenever} \quad x \in \partial E.$$

Here $c' = c'(p, n, \alpha, r_0)$. Using this note in (4.14) we conclude that to prove Lemma 4.3 it suffices to show for some $c'' = c''(p, n, \alpha, r_0) \geq 1$ that

$$\text{(4.15)}\qquad u(y) \leq 1 - d(y, \partial E)/c'' \quad \text{whenever} \quad 0 < d(y, \partial E) < 1/10.$$

To prove (4.15) choose $w \in \partial E$ with $|y - w| = d(y, \partial E)$ and let $\hat{w} := w + \frac{y-w}{|y-w|}$. Then $w \in \partial B(\hat{w}, 1)$, $y \in B(\hat{w}, 1)$, and $E \cap \bar{B}(\hat{w}, 1) = w$, thanks to the convexity of E. From Lemma 4.2 (b) and Harnack's inequality for $1 - u$, we deduce for some $\delta = \delta(p, n, \alpha, r_0)$ with $0 < \delta < 1$ that

$$\text{(4.16)}\qquad (1 - u)(x) \geq \delta \quad \text{whenever} \quad x \in \bar{B}(\hat{w}, 1/2).$$

Using (4.16) and a barrier-type argument as in [**LLN**, Section 2] or [**ALV**, Section 4], it now follows that there exists c_+, depending only on α, p, n, with

$$\text{(4.17)}\qquad c_+\,(1 - u)(x) \geq \delta\, d(x, \partial B(\hat{w}, 1)) \quad \text{whenever} \quad x \in B(\hat{w}, 1).$$

For the readers convenience we outline the proof of (4.17) in the Appendix 7.1. Taking $y = x$ in (4.17) we get (4.15). Thus (4.13) holds so letting $\lambda \to 1$ and using smoothness of ∇u (see Lemma 3.2) and the chain rule we obtain (4.12) (a).

From (4.12) (a), (4.4) (b), and the fact that $\mathrm{Cap}_{\mathcal{A}}(E) \approx c(p, n, \alpha, r_0)$, we obtain,

$$c^{-2}|x|^{\frac{1-n}{p-1}} \le c^{-1}\, u(x)/|x| \le \langle \nabla u(x), -x/|x| \rangle \le |\nabla u(x)|.$$

The right-hand inequality in (4.12) (b) follows from (3.2) $(\hat{a})$, (4.4) (b) and the above fact. □

Before stating our last lemma in this chapter recall from Lemma 4.1 that u is continuous on $\mathbb{R}^n$ with $u \equiv 1$ on E.

LEMMA 4.4. *For each $t \in (0,1)$, the set $\{x \in \mathbb{R}^n : u(x) > t\}$ is convex.*

PROOF. We first prove Lemma 4.4 under the added assumptions that

$$\text{(4.18)} \qquad \mathcal{A} \text{ satisfies (2.8) } (i) \text{ and there exists } r_0 > 0,\ z \in E \text{ with } B(z, r_0) \subset E.$$

Assuming (4.18) we note from Lemma 4.3 and (3.3) that

$$\text{(4.19)} \qquad \begin{cases} |\nabla u| \neq 0 \text{ in } \Omega, \\ u \text{ has Hölder continuous second partials on compact subsets of } \Omega. \end{cases}$$

Our proof of Lemma 4.4 is by contradiction. We follow the proof in [**L**, section 4], although we shall modify it slightly for later ease of use in proving (2.7). We first define for $\hat{x} \in \mathbb{R}^n$,

$$\mathbf{u}(\hat{x}) = \sup \left\{ \min\{u(\hat{y}), u(\hat{z})\}\,; \begin{array}{l} \hat{x} = \lambda \hat{y} + (1-\lambda)\hat{z}, \\ \lambda \in [0,1], \hat{y}, \hat{z} \in \mathbb{R}^n \end{array} \right\}.$$

If Lemma 4.4 is false, then from convexity of E, continuity of u, and the fact that $u(w) \to 0$ as $w \to \infty$, we see there exists $\lambda \in (0,1), \epsilon > 0$, and $x_0 \in \Omega$ such that

$$\text{(4.20)} \qquad 0 < \mathbf{u}^{1+\epsilon}(x_0) - u(x_0) = \max_{\mathbb{R}^n} (\mathbf{u}^{1+\epsilon} - u).$$

For ease of writing we put $\mathbf{v} := \mathbf{u}^{1+\epsilon}$ and $v := u^{1+\epsilon}$. With $\lambda \in (0,1)$ now fixed it follows from the definition of $\mathbf{u}$ that there exists $y_0, z_0 \in \Omega \setminus \{x_0\}$ with

$$\text{(4.21)} \qquad x_0 = \lambda y_0 + (1-\lambda) z_0 \quad \text{and} \quad \mathbf{v}(x_0) = \min\{v(y_0), v(z_0)\}.$$

We first show that

$$\text{(4.22)} \qquad v(y_0) = v(z_0).$$

If, for example, $v(y_0) < v(z_0)$, then since $\nabla u \neq 0$ in Ω, we could choose y' near y_0 with $v(y') > v(y_0)$ and then choose z' so that $(\lambda - 1)(z' - z_0) = \lambda(y' - y_0)$ and $v(z') > v(y')$. Then by similar triangles or algebra, we see first from (4.21) that $x_0 = \lambda y' + (1-\lambda) z'$ and second by construction that

$$\min\{v(y'), v(z')\} > \mathbf{v}(x_0) \text{ which is a contradiction with (4.20).}$$

Thus (4.22) is true. Next we prove that

$$\text{(4.23)} \qquad \xi = \frac{\nabla v(y_0)}{|\nabla v(y_0)|} = \frac{\nabla v(z_0)}{|\nabla v(z_0)|} = \frac{\nabla u(x_0)}{|\nabla u(x_0)|}.$$

Indeed,

$$\frac{\nabla v(y_0)}{|\nabla v(y_0)|} = \frac{\nabla v(z_0)}{|\nabla v(z_0)|}$$

since otherwise we could find y', z' as above with $v(y') > v(y_0), v(z') > v(z_0)$. As previously, we then get a contradiction to (4.20). Finally, armed with this knowledge we see that if (4.23) is false, then we could choose $\nu \in \mathbb{R}^n, |\nu|$ small so that $\mathbf{v}$ is increasing at y_0, z_0 in the direction ν while u is decreasing at x_0 in this direction. Choosing x', y', z' appropriately on rays with direction ν through x_0, y_0, z_0, respectively we again arrive at a contradiction to (4.20). Hence (4.23) is valid.

To simplify our notation, let

$$A = |\nabla v(y_0)|, \ B = |\nabla v(z_0)|, \ C = |\nabla u(x_0)|, \ a = |x_0 - y_0|, \ b = |x_0 - z_0|.$$

From (4.19), we can write

$$\begin{aligned} v(y_0 + \rho\eta) &= v(y_0) + A_1\rho + A_2\rho^2 + o(\rho^2), \\ v(z_0 + \rho\eta) &= v(z_0) + B_1\rho + B_2\rho^2 + o(\rho^2), \\ u(x_0 + \rho\eta) &= u(x_0) + C_1\rho + C_2\rho^2 + o(\rho^2) \end{aligned} \tag{4.24}$$

as $\rho \to 0$ whenever $\langle \xi, \eta \rangle > 0$ for a given $\eta \in \mathbb{S}^{n-1}$. Also

$$A_1/A = B_1/B = C_1/C = \langle \xi, \eta \rangle$$

where the coefficients and $o(\rho^2)$ depend on η. Given η with $\langle \xi, \eta \rangle > 0$ and ρ_1 sufficiently small we see from (4.19) that the inverse function theorem can be used to obtain ρ_2 with

$$v\left(y_0 + \frac{\rho_1}{A}\eta\right) = v\left(z_0 + \frac{\rho_2}{B}\eta\right).$$

We conclude as $\rho_1 \to 0$ that

$$\rho_2 = \rho_1 + \frac{B}{B_1}\left(\frac{A_2}{A^2} - \frac{B_2}{B^2}\right)\rho_1^2 + o(\rho_1^2). \tag{4.25}$$

Now from geometry we see that $\lambda = \frac{b}{a+b}$ so

$$x = x_0 + \eta\frac{[\rho_1\frac{b}{A} + \rho_2\frac{a}{B}]}{a+b} = \lambda(y_0 + \frac{\rho_1}{A}\eta) + (1-\lambda)(z_0 + \frac{\rho_2}{B}\eta).$$

From this equality, (4.25), and Taylor's theorem for second derivatives we have

$$\begin{aligned} u(x) - u(x_0) =& C_1\left[\rho_1\frac{\lambda}{A} + \rho_2\frac{(1-\lambda)}{B}\right] + C_2\left[\rho_1\frac{\lambda}{A} + \rho_2\frac{(1-\lambda)}{B}\right]^2 \\ =& C_1\rho_1\frac{(1-\lambda)A + \lambda B}{AB} + C_1\frac{(1-\lambda)}{B_1}\left(\frac{A_2}{A^2} - \frac{B_2}{B^2}\right)\rho_1^2 \\ &+ C_2\rho_1^2\left(\frac{(1-\lambda)A + \lambda B)}{AB}\right)^2 + o(\rho_1^2). \end{aligned} \tag{4.26}$$

From (4.20) we also have

$$v(y_0 + \frac{\rho_1}{A}\eta) - u(x) \le \mathbf{v}(x) - u(x) \le \mathbf{v}(x_0) - u(x_0) = v(y_0) - u(x_0).$$

Hence the mapping

$$\rho_1 \to v(y_0 + \frac{\rho_1}{A}\eta) - u(x)$$

has a maximum at $\rho_1 = 0$. Using the Taylor expansion for $v(y_0 + \frac{\rho_1}{A}\eta)$ in (4.24) and $u(x)$ in (4.26) we have

$$\begin{aligned} v(y_0 + \frac{\rho_1}{A}\eta) - u(x) =& v(y_0) + \frac{A_1}{A}\rho_1 + \frac{A_2}{A^2}\rho_1^2 - u(x_0) \\ & - C_1\rho_1 \frac{(1-\lambda)A + \lambda B}{AB} - \frac{C_1}{a+b} \frac{a}{B_1}\left(\frac{A_2}{A^2} - \frac{B_2}{B^2}\right)\rho_1^2 \\ & - C_2\rho_1^2 \left(\frac{(1-\lambda)A + \lambda B}{AB}\right)^2 + o(\rho_1^2). \end{aligned}$$

Now from the calculus second derivative test, the coefficient of ρ_1 should be zero and the coefficient of ρ_1^2 should be non-positive. Hence combining terms we get

$$\frac{A_1}{A} = C_1 \frac{(1-\lambda)A + \lambda B}{AB}$$

so taking $\eta = \xi$ we arrive first at

$$\frac{1}{C} = \frac{(1-\lambda)A + \lambda B}{AB} = \frac{(1-\lambda)}{B} + \frac{\lambda}{A}. \tag{4.27}$$

Second, using (4.27) in the ρ_1^2 term we find that

$$0 \geq \frac{A_2}{A^2} - C_1 \frac{(1-\lambda)}{B_1}\left(\frac{A_2}{A^2} - \frac{B_2}{B^2}\right) - \frac{C_2}{C^2}. \tag{4.28}$$

Using $C_1/B_1 = C/B$ and doing some algebra in (4.28) we obtain

$$0 \geq (1-K)\frac{A_2}{A^2} + K\frac{B_2}{B^2} - \frac{C_2}{C^2} \tag{4.29}$$

where

$$K = \frac{(1-\lambda)A}{(1-\lambda)A + \lambda B} < 1.$$

We now focus on (4.29) by writing A_1, B_1, C_1 in terms of derivatives of u and v;

$$0 \geq \sum_{i,j=1}^{n} \left[\frac{(1-K)}{A^2} v_{x_i x_j}(y_0) + \frac{K}{B^2} v_{x_i x_j}(z_0) - \frac{1}{C^2} u_{x_i x_j}(x_0)\right] \eta_i \eta_j. \tag{4.30}$$

From symmetry and continuity considerations we observe that (4.30) holds whenever $\eta \in \mathbb{S}^{n-1}$. Thus, if

$$w(x) = -\frac{(1-K)}{A^2} v(y_0 + x) - \frac{K}{B^2} v(z_0 + x) + \frac{1}{C^2} u(x_0 + x),$$

then the Hessian matrix of w at $x = 0$ is positive semi-definite. That is, $(w_{x_i x_j}(0))$ has non-negative eigenvalues. Also from (i) of Definition 2.1 we see that if

$$a_{ij} = \frac{1}{2}\left[\frac{\partial \mathcal{A}_i}{\partial \eta_j}(\xi)) + \frac{\partial \mathcal{A}_j}{\partial \eta_i}(\xi))\right] \quad \text{for } 1 \leq i, j \leq n,$$

then (a_{ij}) is positive definite. From these two observations we conclude that

$$\operatorname{trace}\left(((a_{ij}) \cdot (w_{x_i x_j}(0))\right) \geq 0. \tag{4.31}$$

To obtain a contradiction we observe from (2.2), the divergence theorem, (4.29), and $p-2$ homogeneity of partial derivatives of $\mathcal{A}_i$, that

$$\sum_{i,j=1}^{n} a_{ij}\, u_{x_j x_i} = |\nabla u|^{2-p} \sum_{i,j=1}^{n} \frac{\partial \mathcal{A}_i}{\partial \eta_j}(\nabla u) u_{x_j x_i} = 0 \quad \text{at } x_0, y_0, z_0. \tag{4.32}$$

Moreover, from the definition of v we have

$$v_{x_i} = (u^{1+\epsilon})_{x_i} = (1+\epsilon)u^{\epsilon}u_{x_i}, \tag{4.33}$$
$$v_{x_ix_j} = (1+\epsilon)\epsilon u^{\epsilon-1}u_{x_i}u_{x_j} + (1+\epsilon)u^{\epsilon}u_{x_ix_j}.$$

Using Definition 2.1, (4.23), and $\mathcal{A}$-harmonicity of u at those points, we find that

(4.34)

$$|\nabla u|^{p-2}\sum_{i,j=1}^{n} a_{ij}\, v_{x_jx_i} = \sum_{i,j=1}^{n}\frac{\partial \mathcal{A}_i}{\partial \eta_j}(\nabla u)[(1+\epsilon)u^{\epsilon-1}u_{x_j}u_{x_i} + (1+\epsilon)u^{\epsilon}u_{x_jx_i}]$$
$$= (1+\epsilon)\epsilon u^{\epsilon-1}\sum_{i,j=1}^{n}\frac{\partial \mathcal{A}_i}{\partial \eta_j}(\nabla u)u_{x_j}u_{x_i} + (1+\epsilon)u^{\epsilon}\sum_{i,j=1}^{n}\frac{\partial \mathcal{A}_i}{\partial \eta_j}(\nabla u)u_{x_jx_i}$$
$$\geq \alpha^{-1}(1+\epsilon)\epsilon u^{\epsilon-1}|\nabla u|^{p-2}|\nabla u|^2 + 0 > 0$$

at points y_0 and z_0 (∇u is also evaluated at these points). Using (4.32), (4.34), we conclude that

$$\operatorname{trace}\left((a_{ij})\cdot(w_{x_ix_j}(0))\right) = \sum_{i,j=1}^{n} a_{ij}\, w_{x_ix_j}(0) < 0. \tag{4.35}$$

Now (4.35) and (4.31) contradict each other. Thus Lemma 4.4 is true when (4.18) holds.

To remove assumption (4.18), suppose $\{\mathcal{A}^{(l)}\}, l = 1, 2, \cdots \in M_p(\alpha/2)$, with

$$\left\{\mathcal{A}^{(l)}, \frac{\partial \mathcal{A}^{(l)}}{\partial \eta_k}\right\} \to \left\{\mathcal{A}, \frac{\partial \mathcal{A}}{\partial \eta_k}\right\} \text{ as } l\to\infty \quad \text{for each } k = 1, 2, \dots, n,$$
$$\text{uniformly on compact subsets of } \mathbb{R}^n\setminus\{0\}.$$

Also assume that (2.8) (i) holds for each l where $\Lambda = \Lambda(l)$. Let

$$E_l = \{x : d(x, E) \leq 1/l\},\ l = 1, 2, \dots,$$

and let u_l be the $\mathcal{A}^{(l)}$-capacitary function corresponding to E_l. From Lemmas 3.1-3.3 and Lemma 4.1, we deduce that a subsequence of $\{u_l\}$ say $\{u_l'\}$ can be chosen so that

$$\{u_l', \nabla u_l'\} \to \{u, \nabla u\} \text{ converges uniformly as } l\to\infty$$
$$\text{on compact subsets of } \mathbb{R}^n \text{ and } \Omega \text{ respectively.}$$

Now from our previous work we see that Lemma 4.4 holds for u_l so

$$E_l(t) = \{x : u_l'(x) > t\} \text{ is convex for } l = 1, 2, \dots, \text{ and } t\in(0,1).$$

Also from Lemma 4.2 these sets are uniformly bounded for a fixed $t \in (0, 1)$. Using these facts, it is easily seen that

$$E(t) = \{x : u(x) > t\} \text{ is convex.}$$

Indeed, if $x, y \in E(t)$ and $t > 4\delta > 0$, then from convexity of $E_l(t)$ and uniform convergence of $\{u_l'\}$ to u we see that the line segment from x to y is contained in $E(t-\delta)$ whenever $2\delta < t$. Letting $\delta \to 0$ we get convexity of $E(t)$.

To prove existence of $\{\mathcal{A}^{(l)}\}$ let

$$\psi(\eta) = \mathcal{A}(\eta/|\eta|) \quad \text{whenever } \eta \in \mathbb{R}^n\setminus\{0\}.$$

Given $\epsilon > 0$, small we also define

$$\psi_\epsilon(\eta) = (\psi * \phi_\epsilon)(\eta) \quad \text{on} \quad B(0,2) \setminus B(0,1/2)$$

where $*$ denotes convolution on $\mathbb{R}^n$ with each component of ψ. Also, $\phi_\epsilon(\eta) = \epsilon^{-n}\phi(\eta/\epsilon)$, and $0 \leq \phi \in C_0^\infty(\mathbb{R}^n)$ with $\int_{\mathbb{R}^n} \phi \, dx = 1$. Set

$$\mathcal{A}_\epsilon(\eta) = |\eta|^{p-1}\psi_\epsilon(\eta/|\eta|) \quad \text{whenever} \quad \eta \in \mathbb{R}^n \setminus \{0\}.$$

Then for ϵ small enough we deduce from Definition 2.1 that $\mathcal{A}_\epsilon \in M_p(\alpha/2)$ and (2.8) (i) holds for $\mathcal{A}_\epsilon$. Letting $\mathcal{A}^{(l)} = \mathcal{A}_{\epsilon_l}$ for sufficiently small ϵ_l with $\epsilon_l \to 0$ we get the above sequence. The proof of Lemma 4.4 is now complete. □

CHAPTER 5

Proof of Theorem A

In the proof of (2.7) we shall need the following lemma.

LEMMA 5.1. *Given $\mathcal{A} \in M_p(\alpha)$, there exists an $\mathcal{A}$-harmonic function G on $\mathbb{R}^n \setminus \{0\}$ and $c = c(p, n, \alpha)$ satisfying*

(5.1)

(*a*) $c^{-1}\,|x|^{(p-n)/(p-1)} \leq G(x) \leq c\,|x|^{(p-n)/(p-1)}$ *whenever* $x \in \mathbb{R}^n \setminus \{0\}$.

(*b*) $c^{-1}\,|x|^{(1-n)/(p-1)} \leq |\nabla G| \leq c\,|x|^{(1-n)/(p-1)}$ *whenever* $x \in \mathbb{R}^n \setminus \{0\}$.

(*c*) *If* $\theta \in C_0^\infty(\mathbb{R}^n)$ *then* $\theta(0) = \int_{\mathbb{R}^n\setminus\{0\}} \langle \mathcal{A}(\nabla G), \nabla\theta\rangle\, dx.$

(*d*) *G is the unique $\mathcal{A}$-harmonic function on $\mathbb{R}^n \setminus \{0\}$ satisfying* (*a*) *and* (*c*).

(*e*) $G(x) = |x|^{(p-n)/(p-1)}\, G(x/|x|)$ *whenever* $x \in \mathbb{R}^n \setminus \{0\}$.

PROOF. Let $\breve{u}$ be the $\mathcal{A}$-capacitary function for $\bar{B}(0,1)$ and let $\breve{\mu}$ be the corresponding capacitary measure for $\bar{B}(0,1)$. Then from (2.6) and Lemma 4.2 we have

$$\breve{\mu}(\bar{B}(0,1)) = \mathrm{Cap}_{\mathcal{A}}(\bar{B}(0,1)) = c_1. \tag{5.2}$$

For $k = 1, 2, \ldots$, let

$$\begin{aligned}
\breve{u}_k(x) &:= c_1^{-\frac{1}{p-1}}\, k^{\frac{n-p}{p-1}}\, \breve{u}(kx) \quad \text{whenever } x \in \mathbb{R}^n,\\
\breve{\mu}_k(F) &:= c_1^{-1}\, k^{n-p} \breve{\mu}(kF) \quad \text{whenever } F \subset \mathbb{R}^n \text{ is a Borel set.}
\end{aligned}$$

Then from Remark 2.3 and Lemma 4.1 we see that $\breve{u}_k$ is continuous on $\mathbb{R}^n$ and $\mathcal{A}$-harmonic in $\mathbb{R}^n \setminus \bar{B}(0,1/k)$ with

$$\breve{u}_k = c_1^{-\frac{1}{p-1}} k^{\frac{n-p}{p-1}} \quad \text{on } \bar{B}(0,1/k).$$

Also if $\phi \in C_0^\infty(\mathbb{R}^n)$ and $\phi_k(x) = \phi(kx)$, then from (3.6) $(i), p-1$ homogeneity of $\mathcal{A}$, and the change of variables theorem, we have

$$\int_{\mathbb{R}^n} \langle \mathcal{A}(\nabla \breve{u}_k), \nabla\phi_k\rangle dx = c_1^{-1} \int_{\mathbb{R}^n} \langle \mathcal{A}(\nabla\breve{u}), \nabla\phi\rangle dx = \int_{\mathbb{R}^n} \phi_k d\breve{\mu}_k. \tag{5.3}$$

Thus $\breve{\mu}_k$ is the measure corresponding to $\breve{u}_k$ with support $\subset \bar{B}(0,1/k)$ and $\breve{\mu}_k(\bar{B}(0,1/k)) = 1$ thanks to (5.2). Also applying (4.4) (*b*) and (4.12) (*b*) to $\breve{u}$ we deduce that

$$\begin{aligned}
&(+) \quad c_+^{-1}\,|x|^{(p-n)/(p-1)} \leq \breve{u}_k(x) \leq c_+|x|^{(p-n)/(p-1)},\\
&(++) \quad c_+^{-1}\,|x|^{(1-n)/(p-1)} \leq |\nabla\breve{u}_k(x)| \leq c_+|x|^{(1-n)/(p-1)}
\end{aligned} \tag{5.4}$$

whenever $|x| \geq 2/k$. Using (5.4), Definition 2.1, and Hölder's inequality, we see that for $\rho > 1/k$,

$$\begin{aligned}\int_{B(0,\rho)} |\mathcal{A}(\nabla \breve{u}_k)| dx &\leq c\, k^{-n/p} \left(\int_{B(0,\rho)\cap B(0,2/k)} |\nabla \breve{u}_k|^p dx \right)^{1-1/p} \\ &\quad + c \int_{B(0,\rho)\setminus B(0,2/k)} |x|^{1-n}\, dx \\ &\leq c^2(k^{-1} + \rho).\end{aligned} \tag{5.5}$$

If $\rho \leq 1/k$, the far right-hand integral is 0 so (5.5) continues to hold. Using Lemmas 3.1, 3.2, we see there is a subsequence of $\{\breve{u}_k\}$ say $\{\breve{u}'_k\}$ with

$$\{\breve{u}'_k, \nabla \breve{u}'_k\} \to \{G, \nabla G\} \text{ converges uniformly as } k \to \infty \text{ on compact subsets of } \mathbb{R}^n \setminus \{0\}.$$

It follows that G is $\mathcal{A}$-harmonic in $\mathbb{R}^n \setminus \{0\}$ and if $\bar{\mu}$ is the measure with mass 1 and support at the origin, then

$$\breve{\mu}_k \rightharpoonup \bar{\mu} \text{ weakly as measures as } k \to \infty.$$

Finally, (5.5) and the above facts imply the sequence $\{|\mathcal{A}(\nabla \breve{u}_k)|\}_{k\geq 1}$ is uniformly integrable on $B(0,\rho)$, so using uniform convergence we get for $\theta \in C_0^\infty(B(0,\rho))$ that

$$\int_{\mathbb{R}^n} \langle \mathcal{A}(\nabla G), \nabla \theta \rangle dx = \lim_{k\to\infty} \int_{\mathbb{R}^n} \langle \mathcal{A}(\nabla \breve{u}_k), \nabla \theta \rangle dx = \lim_{k\to\infty} \int_{\mathbb{R}^n} \theta\, d\breve{\mu}_k = \theta(0)$$

where we have also used (5.3).

To prove uniqueness, suppose v is $\mathcal{A}$-harmonic in $\mathbb{R}^n \setminus \{0\}$ and (a), (c) of (5.1) hold for v and some constant ≥ 1. Observe from (5.1) (a) and (3.2) $(\hat{a})$ that

$$|\nabla v(x)| \leq c^* |x|^{(1-n)/(p-1)} \quad \text{whenever } x \in \mathbb{R}^n \setminus \{0\}. \tag{5.6}$$

Given $\gamma > 0$, let

$$e(x) := G(x) - \gamma v(x) \quad \text{whenever } x \in \mathbb{R}^n \setminus \{0\}.$$

We note that if $\vartheta, \upsilon \in \mathbb{R}^n \setminus \{0\}$ then

$$\mathcal{A}_i(\vartheta) - \mathcal{A}_i(\upsilon) = \sum_{j=1}^n (\vartheta_j - \upsilon_j) \int_0^1 \frac{\partial \mathcal{A}_i}{\partial \eta_j}(t\vartheta + (1-t)\upsilon) dt \tag{5.7}$$

for $i \in \{1, .., n\}$. Using this note it follows that e is a weak solution to

$$\hat{\mathcal{L}} e := \sum_{i,j=1}^n \frac{\partial}{\partial y_i} \left(\hat{a}_{ij}(y) \frac{\partial e}{\partial y_j} \right) = 0 \text{ in } \mathbb{R}^n \setminus \{0\}$$

where

$$\hat{a}_{ij}(y) = \int_0^1 \frac{\partial \mathcal{A}_i}{\partial \eta_j} (t\nabla G(y) + \gamma(1-t)\nabla v(y))\, dt.$$

Moreover from Definition 2.1 (i) we see for some $c = c(p, n, \alpha) \geq 1$ that

$$c^{-1} \sigma(y)\, |\xi|^2 \leq \sum_{i,j=1}^n \hat{a}_{ij}(y) \xi_i \xi_j \quad \text{and} \quad \sum_{i,j=1}^n |\hat{a}_{ij}(y)| \leq c\, \sigma(y) \tag{5.8}$$

whenever $\xi \in \mathbb{R}^n \setminus \{0\}$, where

$$\sigma(y) = \int_0^1 |t\nabla G(y) + (1-t)\gamma \nabla v(y)|^{p-2} dt.$$

Using (5.1) (b) and (5.6) we obtain for $y \in \mathbb{R}^n \setminus \{0\}$ that

$$(5.9) \quad c(\gamma)^{-1} |y|^{\frac{(1-n)(p-2)}{p-1}} \leq \sigma(y) \approx (|\nabla G(y)| + \gamma |\nabla v(y)|)^{p-2} \leq c(\gamma) |y|^{\frac{(1-n)(p-2)}{p-1}}.$$

Here $c(\gamma)$ depends only on γ and those for G, v in (5.1) (a). Thus $G - \gamma v$ is a solution to a linear uniformly elliptic PDE on $B(x, |x|/2)$ whenever $x \in \mathbb{R}^n \setminus \{0\}$ with ellipticity constants that are independent of $x \in \mathbb{R}^n$. Next we observe from (5.1) (a) and the maximum principle for $\mathcal{A}$-harmonic functions that for $r > 0$,

$$(5.10) \qquad \max_{\mathbb{R}^n \setminus B(0,r)} \frac{G}{v} = \max_{\partial B(0,r)} \frac{G}{v} \quad \text{and} \quad \min_{\mathbb{R}^n \setminus B(0,r)} \frac{G}{v} = \min_{\partial B(0,r)} \frac{G}{v}.$$

To continue the proof of (5.1) (d), let

$$\gamma = \liminf_{x \to 0} \frac{G(x)}{v(x)}.$$

Then from (5.10) we see that $G - \gamma v \geq 0$ in $\mathbb{R}^n \setminus \{0\}$ and there exists a sequence $\{z_m\}_{m \geq 1}$ with

$$\lim_{m \to \infty} z_m = (0, \ldots, 0) \text{ and } G(z_m) - \gamma v(z_m) = o(v(z_m)) \text{ as } m \to \infty.$$

Now from Harnack's inequality for linear elliptic PDE and the usual chaining-type argument in balls $B(x, r/2), |x| = r$, we see for some $c' \geq 1$, independent of x, that

$$\max_{\partial B(0,r)} (G - \gamma v) \leq c' \min_{\partial B(0,r)} (G - \gamma v).$$

Using this inequality with $r = |z_m|$, the above facts, and Harnack's inequality for v, we deduce

$$G(x) - \gamma v(x) = o(v(x)) \quad \text{when } |x| = |z_m|.$$

This equality yields in view of (5.10) that first

$$\limsup_{x \to 0} \frac{G(x)}{v(x)} = \gamma$$

and second that $G = \gamma v$. From (5.1) (c) we have $\gamma = 1$ so (5.1) (d) is true.

To prove (5.1) (e) we observe from Remark 2.3 for fixed $t > 0$ that

$$v(x) = t^{(n-p)/(p-1)} G(tx)$$

is $\mathcal{A}$-harmonic in $\mathbb{R}^n \setminus \{0\}$. Also it is easily checked that (5.1) (a) – (c) are valid with G replaced by v. From (5.1) (d) it follows that $G = v$ and thereupon using $t = |x|^{-1}$ that (5.1) (e) is valid. □

We call G the *fundamental solution* or *Green's function* for $\mathcal{A}$-harmonic functions with pole at $(0, \ldots, 0)$. In this chapter we assume only that $E \subset \mathbb{R}^n$ is a compact convex set with $\text{Cap}_{\mathcal{A}}(E) > 0$, in contrast to chapter 4, where we also assumed that $\text{diam}(E) = 1$ and $0 \in E$. Using Lemma 5.1 we prove

LEMMA 5.2. *If u is the $\mathcal{A}$-capacitary function for E and G is as in Lemma* 5.1 *then*

$$\lim_{x \to \infty} \frac{u(x)}{G(x)} = \text{Cap}_{\mathcal{A}}(E)^{\frac{1}{p-1}}.$$

PROOF. Translating and dilating E we see from Remark 2.3 and Lemma 4.2 that there exists, $R_0 = R_0(E, p, n, \alpha) > 100$, such that $E \subset B(0, R_0)$ and

$$c^{-1}\,|x|^{(p-n)/(p-1)} \leq u(x) \leq c\,|x|^{(p-n)/(p-1)} \text{ whenever } |x| \geq R_0$$

where $c = c(E, p, n, \alpha)$. Let $\{R_k\}_{k\geq 1}$ be a sequence of positive numbers $\geq R_0$ with $\lim_{k\to\infty} R_k = \infty$. Put

$$\hat{u}_k(x) = R_k^{\left(\frac{n-p}{p-1}\right)}\ \mathrm{Cap}_{\mathcal{A}}(E)^{-\frac{1}{p-1}}\ u(R_k x) \quad \text{whenever } x \in \mathbb{R}^n,$$

and let $\hat{\mu}_k$ be the measure corresponding to $1 - \hat{u}_k$. Then as in (5.3) we see that $\hat{\mu}_k(\mathbb{R}^n) = 1$ and the support of $\hat{\mu}_k$ is contained in $B(0, R_0/R_k)$. Now arguing as in the proof of (5.1) (c) we get a subsequence of $\{\hat{u}_k\}$ say $\{\hat{u}'_k\}$ with

$$\lim_{k\to\infty} \hat{u}'_k = v \quad \text{uniformly on compact subsets of } \mathbb{R}^n \setminus \{0\}$$

where v is $\mathcal{A}$-harmonic in $\mathbb{R}^n \setminus \{0\}$ and satisfies $(a), (c)$. Thus from (5.1) (d), $v = G$. Since every sequence has a subsequence converging to G we see that

$$\lim_{R\to\infty} R^{\left(\frac{n-p}{p-1}\right)}\ \mathrm{Cap}_{\mathcal{A}}(E)^{-\frac{1}{p-1}}\ u(Rx) = G(x)$$
$$\text{uniformly on compact subsets of } \mathbb{R}^n \setminus \{0\}.$$

Equivalently from (5.1) (e) that

$$\lim_{R\to\infty} \frac{u(Rx)}{G(Rx)} = \mathrm{Cap}_{\mathcal{A}}(E)^{\frac{1}{p-1}} \text{ uniformly on compact subsets of } \mathbb{R}^n \setminus \{0\}.$$

This completes the proof of Lemma 5.2. □

5.1. Proof of (2.7) in Theorem A

In this section we prove, the Brunn-Minkowski inequality for $\mathrm{Cap}_{\mathcal{A}}$, (2.7) in Theorem A.

PROOF OF (2.7). Let E_1, E_2 be as in Theorem A. Put $\Omega_i = \mathbb{R}^n \setminus E_i$ and let u_i be the $\mathcal{A}$-capacitary function for E_i for $i = 1, 2$. Let u be the $\mathcal{A}$-capacitary function for $\lambda E_1 + (1-\lambda)E_2$. Following [**B1, CS**], we note that it suffices to prove

$$\mathrm{Cap}_{\mathcal{A}}(E'_1 + E'_2)^{\frac{1}{n-p}} \geq \mathrm{Cap}_{\mathcal{A}}(E'_1)^{\frac{1}{n-p}} + \mathrm{Cap}_{\mathcal{A}}(E'_2)^{\frac{1}{n-p}} \tag{5.11}$$

whenever E'_i for $i = 1, 2$ are convex sets with $\mathrm{Cap}_{\mathcal{A}}(E'_i) > 0$. To get (2.7) from (5.11) put

$$E'_1 = \lambda E_1 \quad \text{and} \quad E'_2 = (1-\lambda)E_2$$

and use (2.5) (a'). Also to prove (5.11) it suffices to show, for all $\lambda \in (0,1)$ that

$$\mathrm{Cap}_{\mathcal{A}}(\lambda E''_1 + (1-\lambda)E''_2)^{\frac{1}{n-p}} \geq \min\left\{\mathrm{Cap}_{\mathcal{A}}(E''_1)^{\frac{1}{n-p}},\ \mathrm{Cap}_{\mathcal{A}}(E''_2)^{\frac{1}{n-p}}\right\} \tag{5.12}$$

whenever E''_i for $i = 1, 2$ are convex sets with $\mathrm{Cap}_{\mathcal{A}}(E''_i) > 0$. To get (5.11) from (5.12) let

$$E''_i = \mathrm{Cap}_{\mathcal{A}}(E'_i)^{\frac{1}{p-n}}\ E'_i \quad \text{for } i = 1, 2$$

and

$$\lambda = \frac{\mathrm{Cap}_{\mathcal{A}}(E'_1)^{\frac{1}{n-p}}}{\mathrm{Cap}_{\mathcal{A}}(E'_1)^{\frac{1}{n-p}} + \mathrm{Cap}_{\mathcal{A}}(E'_2)^{\frac{1}{n-p}}}$$

then use (2.5) (a') and do some algebra. Thus, we shall only prove (5.12) for E_1, E_2, and all $\lambda \in (0,1)$. Some of our proof is quite similar to the proof of Lemma 4.4.

For this reason we first assume that (4.18) holds for E_1, E_2 and $\mathcal{A}$. Fix $\lambda \in (0,1)$ and set

$$u^*(x) = \sup\left\{\min\{u_1(y), u_2(z)\}\,;\; \begin{array}{c} x = \lambda y + (1-\lambda)z, \\ \lambda \in [0,1], y, z \in \mathbb{R}^n \end{array}\right\}.$$

We claim that

$$u^*(x) \le u(x) \text{ whenever } x \in \mathbb{R}^n. \tag{5.13}$$

Once (5.13) is proved we get (2.7) under assumption (4.18) as follows. From (5.13) and the definition of u^* we have

$$u(x) \ge u^*(x) \ge \min\{u_1(x), u_2(x)\}$$

so from Lemma 5.2

$$\begin{aligned} \mathrm{Cap}_{\mathcal{A}}(\lambda E_1 + (1-\lambda)E_2)^{\frac{1}{p-1}} &= \lim_{|x|\to\infty} \frac{u(x)}{G(x)} \\ &\ge \lim_{|x|\to\infty} \frac{\min\{u_1(x), u_2(x)\}}{G(x)} \\ &= \min\left\{\mathrm{Cap}_{\mathcal{A}}(E_1)^{\frac{1}{p-1}}\,,\, \mathrm{Cap}_{\mathcal{A}}(E_2)^{\frac{1}{p-1}}\right\}. \end{aligned}$$

This finishes proof of (5.12) which implies (5.11) and from our earlier remarks this implies (2.7) in Theorem A under the assumptions (5.13) and (4.18). $\square$

The proof of (5.13) is essentially the same as the proof after (4.18) of Lemma 4.4. Therefore we shall not give all details. From (4.18) we see that $\nabla\hat{u} \neq 0$ and $\hat{u}$ has continuous second partials on $\mathbb{R}^n \setminus \hat{E}$ whenever $\hat{u} \in \{u, u_1, u_2\}$ and $\hat{E} \in \{\lambda E_1 + (1-\lambda)E_2, E_1, E_2\}$. Assume that (5.13) is false. Then there exists $\epsilon > 0$ and $x_0 \in \mathbb{R}^n$ such that if

$$v_1(x) = u_1^{1+\epsilon}(x),\; v_2(x) = u_2^{1+\epsilon}(x), \text{ and } v^*(x) = (u^*)^{1+\epsilon},$$

we have

$$0 < v^*(x_0) - u(x_0) = \max_{\mathbb{R}^n}[\,(u^*)^{1+\epsilon} - u\,]. \tag{5.14}$$

As in (4.20), (4.21), there exists $y_0 \in \Omega_1, z_0 \in \Omega_2$ ($y_0 = x_0 = z_0$ is now possible) with

$$x_0 = \lambda y_0 + (1-\lambda)z_0 \quad \text{and} \quad v^*(x_0) = v_1(y_0) = v_2(z_0).$$

Also as in (4.23) we obtain

$$\xi = \frac{\nabla v_1(y_0)}{|\nabla v_1(y_0)|} = \frac{\nabla v_2(z_0)}{|\nabla v_2(z_0)|} = \frac{\nabla u(x_0)}{|\nabla u(x_0)|} \tag{5.15}$$

so with

$$A = |\nabla v_1(y_0)|,\; B = |\nabla v_2(z_0)|,\; C = |\nabla u(x_0)|,\; a = |x_0 - y_0|,\; b = |x_0 - z_0|.$$

We have

$$\begin{aligned} v_1(y_0 + \rho\eta) &= v_1(y_0) + A_1\rho + A_2\rho^2 + o(\rho^2), \\ v_2(z_0 + \rho\eta) &= v_2(z_0) + B_1\rho + B_2\rho^2 + o(\rho^2), \\ u(x_0 + \rho\eta) &= u(x_0) + C_1\rho + C_2\rho^2 + o(\rho^2) \end{aligned}$$

as $\rho \to 0$ whenever $\langle \xi, \eta\rangle > 0$ and $\eta \in \mathbb{S}^{n-1}$.

We can now essentially copy the argument after (4.24) through (4.35) to eventually arrive at a contradiction to (5.13). Assumption (4.18) for $E_1, E_2, \mathcal{A}$ is removed by the same argument as following (4.18). We omit the details.

CHAPTER 6

Final proof of Theorem A

To prove the statement on equality in the Brunn-Minkowski theorem we shall need the following lemma.

LEMMA 6.1. *Let $\mathcal{A} \in M_p(\alpha)$ satisfy (2.8) (i) and let $E_1, E_2, \Omega_1, \Omega_2, G, u, u_1, u_2$ be as in section 5.1. If*

$$-\frac{G_{\xi\xi}(x)}{|\nabla G(x)|} \geq \tau > 0 \text{ whenever } \xi, x \in \mathbb{S}^{n-1} \text{ with } \langle \nabla G(x), \xi \rangle = 0 \tag{6.1}$$

then there exists $R_1 = R_1(\bar{u}, \alpha, p, n)$, such that if $\bar{u} \in \{u, u_1, u_2\}$, then

$$-\frac{\bar{u}_{\tilde{\xi}\tilde{\xi}}(x)}{|\nabla \bar{u}(x)|} \geq \frac{\tau}{2|x|} > 0 \text{ whenever } \tilde{\xi} \in \mathbb{S}^{n-1},\ |x| > R_1, \text{ with } \langle \nabla \bar{u}(x), \tilde{\xi} \rangle = 0.$$

PROOF. Let $\bar{E} \in \{E_1, E_2, \lambda E_1 + (1-\lambda)E_2\}$ correspond to $\bar{u}$ in Lemma 6.1. We note from Lemma 4.4 that $\{x : \bar{u}(x) \geq 1/2\}$ is convex with nonempty interior and $\min(2\bar{u}, 1)$ is the capacitary function for this set. Thus we can apply Lemmas 4.2, 4.3 to conclude the existence of R_0, and $\bar{c} \geq 1$ depending on the data, $\bar{E}, \bar{u}$ with $\bar{E} \subset B(0, R_0/4)$ and

$$\bar{c}^{-1}|x|^{\frac{1-n}{p-1}} \leq -\langle x/|x|, \nabla \bar{u} \rangle \leq |\nabla \bar{u}|(x) \leq \bar{c}\,|x|^{\frac{1-n}{p-1}} \tag{6.2}$$

whenever $|x| > R_0$. We note that (6.2) also holds for G with $\bar{c}$ replaced by c provided $c = c(p, n, \alpha)$ is large enough, as we see from (4.12) and the construction of G in Lemma 5.1. Set

$$\bar{e}(x) := \bar{u}(x) - \mathrm{Cap}_{\mathcal{A}}(\bar{E})^{\frac{1}{p-1}}\, G(x).$$

From Lemma 5.2 we see that

$$\bar{e}(x) = o(G(x)) = o(|x|^{\frac{p-n}{p-1}}) \quad \text{as } |x| \to \infty. \tag{6.3}$$

Also as in (5.7)-(5.8) we deduce that $\bar{e}$ is a weak solution to the uniformly elliptic *PDE*

$$\bar{\mathcal{L}}\bar{e} := \sum_{i,j=1}^{n} \frac{\partial}{\partial y_i}\left(\bar{a}_{ij}(y)\frac{\partial \bar{e}}{\partial y_j}\right) = 0 \tag{6.4}$$

on $B(x, |x|/2)$ with $|x| \geq R_0$ where

$$\bar{a}_{ij}(y) = \int_0^1 \frac{\partial \mathcal{A}_i}{\partial \eta_j}\left(t\nabla \bar{u}(y) + (1-t)\,\mathrm{Cap}_{\mathcal{A}}(\bar{E})^{\frac{1}{p-1}}\, \nabla G(y)\right) dt.$$

Moreover, for some $c \geq 1$, independent of x, we also have

$$c^{-1}\bar{\sigma}(y)\,|\xi|^2 \leq \sum_{i,j=1}^{n} \bar{a}_{ij}(y)\xi_i\xi_j \text{ and } \sum_{i,j=1}^{n} |\bar{a}_{ij}(y)| \leq c\,\bar{\sigma}(y) \tag{6.5}$$

whenever $\xi \in \mathbb{R}^n \setminus \{0\}$ where $\bar\sigma$ satisfies

$$\bar\sigma(y) \approx (|\nabla \bar u(y)| + \mathrm{Cap}_{\mathcal{A}}(\bar E)^{\frac{1}{p-1}} |\nabla G(y)|)^{p-2} \approx |x|^{\frac{(1-n)(p-2)}{p-1}} \tag{6.6}$$

for $|x| \geq R_0$. Constants depend on various quantities but are independent of x.

From well-known results for uniformly elliptic PDE (see [**GT**]) we see that

$$|x|^{-n/2} \left(\int_{B(x,|x|/4)} |\nabla \bar e|^2 \, dy \right)^{1/2} \leq c\,|x|^{-1} \max_{B(x,|x|/2)} \bar e = o\left(|x|^{\frac{1-n}{p-1}}\right) \text{ as } x \to \infty, \tag{6.7}$$

where c as above depends on various quantities but is independent of x. From (6.7), weak type estimates, and Lemma 3.2 $(\hat a)$ for $\bar u, G$ we also have

$$|\nabla \bar e(x)| = o\left(|x|^{\frac{1-n}{p-1}}\right) \text{ as } x \to \infty. \tag{6.8}$$

Indeed, given $\epsilon > 0$, we see from (6.7) that there exists $\rho = \rho(\epsilon)$ large, such that if $|x| \geq \rho$, then

$$|\nabla \bar e| \leq \epsilon |x|^{\frac{1-n}{p-1}} \text{ on } B(x,|x|/2)$$

except on a set $\Gamma \subset B(x,|x|/2)$ with

$$\mathcal{H}^n(\Gamma) \leq \epsilon^{n+1} |x|^n.$$

If $y \in \Gamma$ and ϵ is small enough there exists $z \in B(x,|x|/2) \setminus \Gamma$ with $|z-y| \leq \epsilon|x|$. Then from (3.2) $(\hat a)$ for $\bar u, G$ we deduce

$$|\nabla \bar e(y)| \leq \epsilon |x|^{\frac{1-n}{p-1}} + |\nabla \bar e(y) - \nabla \bar e(z)| \leq \epsilon^{\beta/2}\, |x|^{\frac{1-n}{p-1}}$$

for ϵ small enough and $|x| \geq \rho$. Since ϵ is arbitrary we conclude the validity of (6.8).

We claim that also,

$$\begin{aligned} |x|^{-n/2} \left(\int_{B(x,|x|/4)} \sum_{i,j=1}^{n} \left| \frac{\partial^2 \bar e}{\partial y_i \partial y_j} \right|^2 dy \right)^{1/2} &\leq c|x|^{-2} \max_{B(x,|x|/2)} \bar e \\ &= o\left(|x|^{\frac{2-n-p}{p-1}}\right) \text{ as } |x| \to \infty. \end{aligned} \tag{6.9}$$

To prove (6.9) we first observe from (6.2) for $\bar u, G$ that

$$\begin{aligned} &\left(t|\nabla \bar u(z)| + (1-t)\, \mathrm{Cap}_{\mathcal{A}}(\bar E)^{\frac{1}{p-1}}\, |\nabla G(z)|\right) \\ &\qquad \leq \left| t\langle \nabla \bar u(z), z/|z|\rangle + (1-t)\, \mathrm{Cap}_{\mathcal{A}}(\bar E)^{\frac{1}{p-1}}\, \langle \nabla G(z), z/|z|\rangle \right| \end{aligned} \tag{6.10}$$

when $z \in B(x,|x|/2)$ and $|x| \geq R_0$. Using (6.10), (2.8) (i), (6.2), and (3.3) for $\bar u, G$ we deduce for some $\breve c \geq 1$ and $\mathcal{H}^n$ almost every $\hat x, \hat y \in B(x,|x|/2)$ with $|\hat x - \hat y| \leq |x|/\breve c$ that

$$\begin{aligned} &|\bar a_{ij}(\hat x) - \bar a_{ij}(\hat y)| \\ &\quad \leq \breve c\, |\hat x - \hat y| \max_{B(x,|x|/2)} \left\{ (|\nabla \bar u(z)| + |\nabla G(z)|)^{p-3} \sum_{i,j=1}^{n} (|u_{z_i z_j}(z)| + |G_{z_i z_j}(z)|) \right\} \\ &\quad \leq \breve c^2 |\hat x - \hat y|\, |x|^{\frac{(2-p)n-1}{p-1}} \end{aligned} \tag{6.11}$$

where $\breve{c}$ is independent of $x, \hat{x}, \hat{y}$ subject to the above requirements. Next we use the method of difference quotients. Recall from the introduction that e_m denotes the point with x_l coordinate $= 0, l \neq m$, and $x_m = 1$. Let

$$q_{h,m}(\hat{y}) = \frac{q(\hat{y} + he_m) - q(\hat{y})}{h}$$

whenever q is defined at $\hat{y}$ where $\hat{y} + he_m \in B(\hat{x}, |x|/\breve{c})$. Let ϕ be a non-negative functions satisfying

$$\phi \in C_0^\infty(B(\hat{x}, |x|/(4\breve{c}))) \quad \text{with} \quad \phi \equiv 1 \text{ on } B(\hat{x}, |x|/(8\breve{c})) \quad \text{and} \quad |\nabla \phi| \leq c_* |x|^{-1}.$$

Choosing appropriate test functions in (6.4) we see for $1 \leq m \leq n$ that

$$0 = \int_{B(\hat{x}, |x|/(4\breve{c}))} \sum_{i,j=1}^{n} (\bar{a}_{ij} \bar{e}_{\hat{y}_i})_{h,m} \, (\bar{e}_{h,m} \, \phi^2)_{\hat{y}_j} d\hat{y}. \tag{6.12}$$

Using (6.5), (6.6), (6.8), (6.11) to make estimates in (6.12), as well as Cauchy's inequality with epsilon, we find for some $c \geq 1$, independent of $x, \hat{x}$,

$$\begin{aligned}
|x|^{\frac{(1-n)(p-2)}{p-1}} & \int_{B(\hat{x}, |x|/(4\breve{c}))} |\nabla \bar{e}_{h,m}|^2 \, \phi^2 \, d\hat{y} \\
& \leq c \int_{B(\hat{x}, |x|/(4\breve{c}))} \sum_{i,j=1}^{n} a_{ij} (\bar{e}_{\hat{y}_i})_{h,m} \, (\bar{e}_{\hat{y}_j}\,)_{h,m} \, \phi^2 \, d\hat{y} \\
& \leq c^2 \int_{B(\hat{x}, |x|/\breve{c})} \sum_{i,j=1}^{n} |(\bar{a}_{ij})_{h,m}| |\bar{e}_{\hat{y}_i}(\hat{y} + h)| \, |(\bar{e}_{h,m} \, \phi^2)_{\hat{y}_j}| d\hat{y} \\
& \quad + \frac{c^2}{|x|} \int_{B(\hat{x}, |x|/(4\breve{c}))} \sum_{i,j=1}^{n} |a_{ij} (\bar{e}_{\hat{y}_i})_{h,m} \, \bar{e}_{\hat{y}_j}| \, \phi \, d\hat{y} \\
& \leq (1/2) |x|^{\frac{(1-n)(p-2)}{p-1}} \int_{B(\hat{x}, |x|/(4\breve{c}))} |\nabla \bar{e}_{h,m}|^2 \, \phi \, d\hat{y} + o\left(|x|^{\frac{2-n-p}{p-1}}\right).
\end{aligned} \tag{6.13}$$

It follows from (6.13) after some algebra that

$$|x|^{-n} \left(\int_{B(\hat{x}, |x|/(8\breve{c}))} |\nabla \bar{e}_{h,m}|^2 \, d\hat{y} \right)^{1/2} = o\left(|x|^{\frac{2-n-p}{p-1}}\right) \text{ as } |x| \to \infty. \tag{6.14}$$

Letting $h \to 0$ in (6.14) and covering $B(x, |x|/2)$ by a finite number of balls of the form $B(\hat{x}, |x|/c)$, we get (6.9). From (6.9), (3.3), and weak type estimates it follows, as in the proof of (6.8), that

$$\sum_{i,j=1}^{n} \left| \frac{\partial^2 \bar{e}}{\partial x_i \partial x_j} \right| = o\left(|x|^{\frac{2-n-p}{p-1}}\right) \text{ as } x \to \infty. \tag{6.15}$$

We omit the details.

We now prove Lemma 6.1. Suppose for some large x and $\tilde{\xi} \in \mathbb{S}^{n-1}$ that $\langle \nabla \bar{u}(x), \tilde{\xi} \rangle = 0$. Then from (6.8) and (6.2) for G we see that

$$\langle \nabla G(x), \xi \rangle = o\left(|x|^{(1-n)/(p-1)}\right) = o\left(|\nabla G(x)|\right) \text{ as } x \to \infty.$$

From this inequality we deduce that $\tilde{\xi} = \xi + \lambda$ where ξ is orthogonal to $\nabla G(x)$ and λ points in the same direction as $\nabla G(x)$ with $|\lambda| = o(1)$ as $x \to \infty$. Using these facts, (6.8), (6.15), (6.1), (3.3) for G, as well as homogeneity of G and its derivatives, we have for large $|x|$,

$$\frac{\bar{u}_{\tilde{\xi}\tilde{\xi}}(x)}{|\nabla \bar{u}(x)|} = (1+o(1))\frac{G_{\tilde{\xi}\tilde{\xi}}(x)}{|\nabla G(x)|} = o(1)|x|^{-1} + \frac{G_{\xi\xi}(x)}{|\nabla G(x)|} \geq (\tau/2)\,|x|^{-1}$$

for $|x| \geq R_0$ provided R_0 is large enough. This finishes the proof of Lemma 6.1. □

Next we state

LEMMA 6.2. *If G is the Green's function for an $\mathcal{A} \in M_p(\alpha)$ satisfying* (2.8) *(ii) then* (6.1) *is valid for some $\tau > 0$.*

The proof of Lemma 6.2 is given in Appendix 7.2. We continue the proof equality in Theorem A under the assumption that Lemma 6.2 is valid. Let u, u_1, u_2, E_1, E_2, be as in Lemma 6.1. Following [**CS**] we note for u^* as in (5.13) that

$$\{u^*(x) \geq t\} = \lambda\{u_1(y) \geq t\} + (1-\lambda)\{u_2(z) \geq t\} \tag{6.16}$$

whenever $t \in (0,1)$. Indeed containment of the left-hand set in the right-hand set is a direct consequence of the definition of u^*. Containment of the right-hand set in the left-hand set follows from the fact that if $u^*(x) = \min\{u_1(y), u_2(z)\}$ for some $y \in E_1, z \in E_2$, with $x = \lambda y + (1-\lambda)z$, then $u_1(y) = u_2(z)$. This fact is proved by the same argument as in the proof of (4.22) or the display below (5.14).

If equality holds in (2.7) in Theorem A for some $\lambda \in (0,1)$, we first observe from Lemma 4.1 and (4.4) (a) that for almost every $t \in (0,1)$,

$$\mathrm{Cap}_{\mathcal{A}}(\{\hat{u} \geq t\}) = t^{1-p}\,\mathrm{Cap}_{\mathcal{A}}(\{\hat{u} \geq 1\}) \text{ whenever } \hat{u} \in \{u_1, u_2, u\}$$

and second that

$$\mathrm{Cap}_{\mathcal{A}}(\{u \geq t\})^{\frac{1}{n-p}} = \lambda\,\mathrm{Cap}_{\mathcal{A}}(\{u_1 \geq t\})^{\frac{1}{n-p}} + (1-\lambda)\,\mathrm{Cap}_{\mathcal{A}}(\{u_2 \geq t\})^{\frac{1}{n-p}}. \tag{6.17}$$

On the other hand, using (6.16), convexity of $\{u_i \geq t\}, i = 1,2$, and (2.7) we obtain

$$\mathrm{Cap}_{\mathcal{A}}(\{u^* \geq t\})^{\frac{1}{n-p}} \geq \lambda\,\mathrm{Cap}_{\mathcal{A}}(\{u_1 \geq t\})^{\frac{1}{n-p}} + (1-\lambda)\,\mathrm{Cap}_{\mathcal{A}}(\{u_2 \geq t\})^{\frac{1}{n-p}}. \tag{6.18}$$

We conclude from (6.17), (6.18) that for almost every $t \in (0,1)$

$$\mathrm{Cap}_{\mathcal{A}}(\{u^* \geq t\}) \geq \mathrm{Cap}_{\mathcal{A}}(\{u \geq t\}). \tag{6.19}$$

Now from (5.13) we see that $u^* \leq u$ so $\{u^* \geq t\} \subset \{u \geq t\}$. This fact and (6.19) imply for almost every $t \in (0,1)$ that

$$\{u^* \geq t\} = \{u \geq t\}. \tag{6.20}$$

To prove this statement let U^*, U be the corresponding $\mathcal{A}$-capacitary functions for these sets. Then from the maximum principle for $\mathcal{A}$-harmonic functions and Lemma 4.1 we see that $U - U^* \geq 0$ in $\mathbb{R}^n$. Moreover, from (4.12) (a) we deduce as in (5.7)-(5.9) that $U - U^*$ satisfies a uniformly elliptic PDE locally in $\mathbb{R}^n \setminus \{x : U(x) \geq 1\}$ for which non-negative solutions satisfy a Harnack inequality. It follows from Harnack's inequality and the usual chaining argument that either

$$(+) \qquad U \equiv U^* \text{ in } \mathbb{R}^n \setminus \{x : U(x) \geq 1\}$$

which implies (6.20), or

$$(++) \qquad U - U^* > 0 \text{ in } \mathbb{R}^n \setminus \{x : U(x) \geq 1\}.$$

If (++) holds we see from a continuity argument that there exists $\rho > 0, \gamma > 1$ for which U, U^* are $\mathcal{A}$-harmonic in $\mathbb{R}^n \setminus \bar{B}(0,\rho)$ and $U/U^* \geq \gamma$ on $\partial B(0,\rho)$. Using the maximum principle for $\mathcal{A}$-harmonic functions it would then follow that

$$U \geq \gamma U^* \text{ in } \mathbb{R}^n \setminus \bar{B}(0,\rho).$$

Dividing this inequality by G and taking limits as in Lemma 5.2 we get, in contradiction to (6.19), that

$$\mathrm{Cap}_{\mathcal{A}}(\{u \geq t\}) > \mathrm{Cap}_{\mathcal{A}}(\{u^* \geq t\}).$$

This proves (6.20). From continuity of u, u^* we conclude first that (6.20) holds for every $t \in (0,1)$ and second that $u^* \equiv u$ in $\mathbb{R}^n$. Thus (6.16) is valid with u^* replaced by u.

For fixed $t \in (0,1)$, let $h_i(\cdot,t)$ be the support function for $\{u_i \geq t\}$ for $i = 1,2$, and let $h(\cdot,t)$ be the support function for $\{u \geq t\}$. More specifically for $X \in \mathbb{R}^n$ and $t \in (0,1)$

$$h_i(X,t) := \sup_{x\in\{u_i\geq t\}} \langle X, x\rangle \text{ for } i = 1,2 \quad\text{and}\quad h(X,t) := \sup_{x\in\{u\geq t\}} \langle X, x\rangle.$$

From (6.16) with u^* replaced by u and the above definitions we see that

$$(6.21)\quad h(X,t) = \lambda h_1(X,t) + (1-\lambda)h_2(X,t) \quad \text{for every } X \in \mathbb{R}^n \text{ and } t \in (0,1).$$

We note from (3.3) and Lemmas 4.3, 4.4, that $\nabla\bar{u} \neq 0$ and $\bar{u}$ has locally Hölder continuous second partials in $\{\bar{u} < 1\}$ whenever $\bar{u} \in \{u_1, u_2, u\}$. From Lemmas 6.1 and 6.2 we see there exists $t_0, \tau_0 > 0$ small and R_0 large such that if $\bar{u} \in \{u_1, u_2, u\}$ then

(6.22)

$$(*) \quad \{\bar{u} \leq t\} \subset \mathbb{R}^n \setminus \bar{B}(0,R_0) \text{ for } t \leq t_0 \leq 1/4,$$

$$(**) \quad -\frac{\bar{u}_{\tilde{\xi}\tilde{\xi}}(x)}{|\nabla\bar{u}(x)|} \geq \tau_0 \quad \text{whenever } \tilde{\xi} \in \mathbb{S}^{n-1} \text{ and } |x| \geq R_0 \text{ with } \langle\nabla\bar{u}(x), \tilde{\xi}\rangle = 0.$$

From (6.22) we see that the curvatures at points on $\{\bar{u} = t\}$ are bounded away from 0 when $t \leq t_0$. Thus

$$-\frac{\nabla\bar{u}}{|\nabla\bar{u}|} \text{ is a 1-1 mapping from } \{\bar{u} = t\} \text{ onto } \mathbb{S}^{n-1}$$

while

$$\left(-\frac{\nabla\bar{u}}{|\nabla\bar{u}|}, \bar{u}\right) \text{ is a 1-1 mapping from } \{u < t_0\} \text{ onto } \mathbb{S}^{n-1} \times (0,t_0).$$

From (6.22), elementary geometry, and the inverse function theorem it follows that if $\bar{h}$ is the support function corresponding to $\bar{u} \in \{u, u_1, u_2\}$ and $0 < t < t_0$, then $\bar{h}$ has Hölder continuous second partials and

$$(6.23)\qquad \nabla_X \bar{h}(X,t) = \bar{x}(X,t)$$

where $\bar{x}$ is the point in $\{\bar{u} = t\}$ with

$$\frac{X}{|X|} = -\frac{\nabla\bar{u}(\bar{x})}{|\nabla\bar{u}(\bar{x})|}.$$

In (6.23), ∇_X denotes the gradient in the X variable only. Also h is homogeneous of degree one in the X variable so (6.23) implies

$$\bar{h}(X,t) = \langle X, \bar{x}(X,t)\rangle \quad \text{and} \quad \frac{\partial \bar{h}}{\partial t} = \langle X, \frac{\partial \bar{x}}{\partial t}\rangle.$$

Since $\bar{u}(\bar{x}) = t$ and $-\nabla \bar{u}(\bar{x})/|\nabla \bar{u}(\bar{x})| = X/|X|$ we get from the chain rule that

$$1 = \langle \nabla \bar{u}, \frac{\partial \bar{x}}{\partial t}\rangle = -|\nabla u(\bar{x})| \, \frac{\partial \bar{h}}{\partial t}. \tag{6.24}$$

Next since $0 = u^* - u$ has an absolute maximum at each $x \in \{u < 1\}$ we can repeat the argument in (4.22), (4.23) to deduce that there exists $y \in \{u_1 < 1\}$, $z \in \{u_2 < 1\}$ with

$$x = \lambda y + (1-\lambda) z \quad \text{and} \quad u(x) = u_1(y) = u_2(z). \tag{6.25}$$

Repeating the argument leading to (4.23) or (5.15) we find that

$$\xi = \frac{\nabla u_1(y)}{|\nabla u_1(y)|} = \frac{\nabla u_2(z)}{|\nabla u_2(z)|} = \frac{\nabla u(x)}{|\nabla u(x)|} \tag{6.26}$$

and after that

$$\begin{aligned} u_1(y+\rho\eta) &= u_1(y) + A_1\rho + A_2\rho^2 + o(\rho^2), \\ u_2(z+\rho\eta) &= u_2(z) + B_1\rho + B_2\rho^2 + o(\rho^2), \\ u(x+\rho\eta) &= u(x) + C_1\rho + C_2\rho^2 + o(\rho^2) \end{aligned} \tag{6.27}$$

as $\rho \to 0$ whenever $\langle \xi, \eta\rangle > 0$ and $\eta \in \mathbb{S}^{n-1}$, where

$$A = |\nabla u_1(y)|, \; B = |\nabla u_2(z)|, \; C = |\nabla u(x)|, \; \lambda = \frac{b}{a+b}.$$

Using (6.27) and once again repeating the argument leading to (4.30) we first arrive at

$$0 \geq \sum_{i,j=1}^{n} \left[\frac{(1-K)}{A^2} (u_1)_{x_i x_j}(y) + \frac{K}{B^2} (u_2)_{x_i x_j}(z) - \frac{1}{C^2} u_{x_i x_j}(x) \right] \eta_i \eta_j \tag{6.28}$$

where as earlier,

$$\frac{1}{C} = \frac{(1-\lambda)A + \lambda B}{AB} = \frac{1-\lambda}{B} + \frac{\lambda}{A} \quad \text{and} \quad K = \frac{(1-\lambda)A}{\lambda B + (1-\lambda)A}. \tag{6.29}$$

Using (6.28) we can argue as below (4.30) to deduce first that if

$$w(\hat{x}) = -\frac{(1-K)}{A^2} u_1(y+\hat{x}) - \frac{K}{B^2} u_2(z+\hat{x}) + \frac{1}{C^2} u(x+\hat{x}),$$

then the Hessian matrix of w at $\hat{x} = 0$ is positive semi-definite. Second if

$$a_{ij} = \frac{1}{2} \left[\frac{\partial \mathcal{A}_i}{\partial \eta_j}(\xi) + \frac{\partial \mathcal{A}_j}{\partial \eta_i}(\xi) \right] \quad \text{for } 1 \leq i, j \leq n,$$

then (a_{ij}) is positive definite and from $\mathcal{A}$-harmonicity of u, u_1, u_2, as well as (6.26),

$$\operatorname{trace}\big((a_{ij}) \cdot (w_{x_i x_j}(0))\big) = 0.$$

From this equality we conclude that the Hessian of w vanishes at $\hat{x} = 0$ so by continuity, equality holds in (6.28) whenever $\eta \in \mathbb{S}^{n-1}$.

Using (6.21) we shall convert this equality into an inequality involving support functions from which we can make conclusions. We shall need the following lemma from [**CS**]:

LEMMA 6.3 ([**CS**], Lemma 2). *Let H_1, H_2, be symmetric positive definite matrices and let $0 < r, s$. Then for every $\lambda \in [0,1]$ the following inequality holds:*

$$(\lambda s+(1-\lambda)r)^2 \operatorname{trace}\left[(\lambda H_1+(1-\lambda)H_2)^{-1}\right] \leq \lambda s^2 \operatorname{trace}\left[H_1^{-1}\right]+(1-\lambda)r^2 \operatorname{trace}\left[H_2^{-1}\right].$$

Equality holds if and only if

$$rH_1 = sH_2.$$

To convert (6.21) into an equality involving support functions we first assume that

$$\xi = e_n = (0,\dots,0,1) \quad \text{and} \quad u(x) = u_1(y) = u_2(z) = t \tag{6.30}$$

in (6.25), (6.26). Then $X/|X| = e_n$ and from (6.23), as well as, 0-homogeneity of the components of $\nabla_X \bar{h}$ we see for fixed $t \in (0,t_0)$ that

$$\bar{h}_{X_k X_n} = 0 \quad \text{for } k = 1,\dots,n.$$

Also from the chain rule we deduce for $1 \leq i,j \leq n-1$ that

$$\delta_{ij} = \sum_{i=1}^{n-1} \bar{h}_{X_i X_k} \frac{\partial X_k}{\partial x_j} = \sum_{i=1}^{n-1} \bar{h}_{X_i X_k} \frac{-\bar{u}_{x_k x_j}}{|\nabla \bar{u}|} \tag{6.31}$$

when $X/|X| = e_n$, where δ_{ij} is the Kronecker δ and partial derivatives of $\bar{u}$ are evaluated at x, y, z, depending on whether $\bar{u} = u, u_1, u_2$, respectively. For $1 \leq i,j \leq n-1$, consider $(\bar{h}_{X_i X_j})$ and $\left(\frac{-\bar{u}_{x_i x_j}}{|\nabla \bar{u}|}\right)$ as $(n-1)\times(n-1)$ matrices. Then (6.31) implies (for $1 \leq i,j \leq n-1$)

$$(\bar{h}_{X_i X_j}) \text{ is the inverse of the positive definite matrix } \left(\frac{-\bar{u}_{x_i x_j}}{|\nabla \bar{u}|}\right). \tag{6.32}$$

For $1 \leq i,j \leq n-1$, let

$$H_k := ((h_k)_{X_i X_j}) \text{ for } k = 1,2 \quad \text{and} \quad H := (h_{X_i X_j})$$

be the $(n-1)\times(n-1)$ matrices of second partials of the support functions corresponding to h_1, h_2, h, respectively. Using $\eta = e_i$ for $1 \leq i \leq n-1$, and multiplying each side of the equality in (6.28) by $AB[(1-\lambda)A + \lambda B]$ we see in view of (6.29), (6.32), after some algebra that the resulting equality can be rewritten in terms of our new notation as

$$(\lambda B + (1-\lambda)A)^2 H^{-1} = \lambda B^2 H_1^{-1} + (1-\lambda)A^2 H_2^{-1}. \tag{6.33}$$

Now from (6.21) we also have $H = \lambda H_1 + (1-\lambda)H_2$ so obviously, $H^{-1} = (\lambda H_1 + (1-\lambda)H_2)^{-1}$. Using this equality in (6.33) we conclude from Lemma 6.3 with $A = r, B = s$ that at (X,t), $AH_1 = BH_2$ and thereupon from (6.29), (6.21) that

$$AH_1 = BH_2 = CH \text{ at } (X,t) \text{ when (6.30) holds.} \tag{6.34}$$

We continue under assumption (6.30). Following [**CS**, page 470], we will compute $\bar{u}_{x_n x_n}(\bar{x})$ in terms of second partial derivatives of $\bar{h}(X,t)$ where $\bar{x} = x, y$, or z in (6.25) depending on whether $\bar{u} = u, u_1$ or u_2. From the chain rule, (6.24), and

(6.30),

$$
\begin{aligned}
-\bar{u}_{x_n x_n} &= \frac{\partial}{\partial x_n}\left(\frac{1}{\bar{h}_t(X,t)}\right) = -\frac{1}{\bar{h}_t^2}\left[\sum_{i=1}^{n}\frac{\partial \bar{h}_t}{\partial X_i}\frac{\partial X_i}{\partial x_n} + \bar{h}_{tt}\frac{\partial t}{\partial x_n}\right] \\
&= -\frac{1}{\bar{h}_t^2}\sum_{i=1}^{n}\frac{\partial \bar{h}_t}{\partial X_i}\frac{\partial X_i}{\partial x_n} + \frac{1}{\bar{h}_t^3}\bar{h}_{tt}.
\end{aligned} \tag{6.35}
$$

Taking derivatives in (6.23) we also have for $i = 1, \ldots, n-1$,

$$
0 = \sum_{j=1}^{n-1}\bar{h}_{X_iX_j}\frac{\partial X_j}{\partial x_n} + \bar{h}_{X_it}\frac{\partial t}{\partial x_n} = \sum_{j=1}^{n-1}\bar{h}_{X_iX_j}\frac{\partial X_j}{\partial x_n} - \frac{\bar{h}_{X_it}}{\bar{h}_t}. \tag{6.36}
$$

Using (6.36) to solve for $\frac{\partial X}{\partial x_n}$ and then putting the result in (6.35) we obtain at $(\bar{x}, t)$,

$$
\begin{aligned}
-\bar{u}_{x_n x_n} &= -\frac{1}{\bar{h}_t^2}\sum_{i=1}^{n-1}\bar{h}_{tX_i}\frac{\partial X_i}{\partial x_n} + \frac{1}{\bar{h}_t^3}\bar{h}_{tt} \\
&= -\frac{1}{\bar{h}_t^3}\left[\langle \nabla_X \bar{h}_t\,(\bar{h}_{X_iX_j})^{-1}, \nabla_X \bar{h}_t\rangle - \bar{h}_{tt}\right]
\end{aligned} \tag{6.37}
$$

where $\nabla_X \bar{h}$ is written as a $1 \times n-1$ row matrix. Let M denote the inverse of the matrix in (6.34). Note that M is positive definite and symmetric. Using (6.34), (6.24), (6.37), as well as the notation used previously for gradients of u, u_1, u_2 at x, y, z, we see that

$$
\begin{aligned}
-\frac{u_{x_nx_n}(x)}{C^2} &= C^2\,\langle \nabla_X h_t\, M, \nabla_X h_t\rangle - C h_{tt}, \\
-\frac{(u_1)_{x_nx_n}(y)}{A^2} &= A^2\,\langle \nabla_X (h_1)_t\, M, \nabla_X (h_1)_t\rangle - A\,(h_1)_{tt}, \\
-\frac{(u_2)_{x_nx_n}(z)}{B^2} &= B^2\,\langle \nabla_X (h_2)_t\, M, \nabla_X (h_2)_t\rangle - B\,(h_2)_{tt}.
\end{aligned} \tag{6.38}
$$

Using (6.38) in the equality in (6.28) with $\eta = e_n$, we find that

$$
\begin{aligned}
&C^2\langle \nabla_X h_t\, M, \nabla_X h_t\rangle - C h_{tt} \\
&\quad = \frac{\lambda B}{\lambda B + (1-\lambda)A}\left[A^2\,\langle \nabla_X (h_1)_t\, M, \nabla_X (h_1)_t\rangle - A(h_1)_{tt}\right] \\
&\qquad + \frac{(1-\lambda)A}{\lambda B + (1-\lambda)A}\left[B^2\langle \nabla_X (h_2)_t\, M, \nabla_X (h_2)_t\rangle - B(h_2)_{tt}\right].
\end{aligned} \tag{6.39}
$$

Since

$$
h = \lambda h_1 + (1-\lambda)h_2 \quad \text{and} \quad C = \frac{AB}{\lambda B + (1-\lambda)A},
$$

the terms involving two derivatives in t on both sides of (6.39) are equal so may be removed. Doing this and using above identity involving h and C once again we arrive at

(6.40)
$$\frac{A^2B^2}{(\lambda B+(1-\lambda)A)^2}\langle(\lambda\nabla_X(h_1)_t+(1-\lambda)\nabla(h_2)_t)\,M,(\lambda\nabla_X(h_1)_t+(1-\lambda)\nabla(h_2)_t)\rangle$$
$$=\frac{\lambda A^2B}{\lambda B+(1-\lambda)A}\langle\nabla_X(h_1)_t\,M,\nabla_X(h_1)_t\rangle + \frac{(1-\lambda)AB^2}{\lambda B+(1-\lambda)A}\langle\nabla_X(h_2)_t\,M,\nabla_X(h_2)_t\rangle.$$

For ease of notation let
$$\Upsilon := \frac{\lambda(1-\lambda)}{(\lambda B+(1-\lambda)A)^2}.$$

Multiplying (6.40) with this expression out, using partial fractions, and gathering terms in $\langle\nabla_X(h_i)_tM,\nabla_X(h_i)_t\rangle$ for $i=1,2$, we see that
$$2\Upsilon A^2B^2\langle\nabla_X(h_1)_tM,\nabla(h_2)_t\rangle$$
$$=\Upsilon A^3B\langle\nabla_X(h_1)_t\,M,\nabla_X(h_1)_t\rangle+\Upsilon AB^3\langle\nabla_X(h_2)_t\,M,\nabla_X(h_2)_t\rangle.$$

This equality can be factored into
$$\Upsilon\langle(A^{3/2}B^{1/2}\nabla_X(h_1)_t-B^{3/2}A^{1/2}\nabla(h_2)_t)\,M,A^{3/2}B^{1/2}\nabla_X(h_1)_t-B^{3/2}A^{1/2}\nabla(h_2)_t\rangle$$
$$=0.$$

Since M is positive definite we conclude from this equality that

(6.41)
$$A\nabla_X(h_1)_t=B\nabla_X(h_2)_t \quad\text{or equivalently that}\quad \nabla_X\log\left(\frac{(h_1)_t}{(h_2)_t}\right)=0.$$

For $i=1,2$, let $\bar{x}_i(X,t)$ be the parametrization of $\{u_i=t\}$ in (6.23) for $t<t_0$ and $X\in\mathbb{S}^{n-1}$. From (6.34), (6.41), and (6.23) we see that if (6.30) holds then

(6.42)
$$\frac{\partial}{\partial X_i}\left(\frac{|\nabla u_2|(\bar{x}_1)}{|\nabla u_1|(\bar{x}_2)}\right)=0 \quad\text{and}\quad \frac{\partial}{\partial X_i}\left(\bar{x}_1-\bar{x}_2\frac{|\nabla u_2|(\bar{x}_1)}{|\nabla u_1|(\bar{x}_2)}\right)=0$$

when $1\le i\le n$ at (e_n,t).

At this point, we remove the assumption (6.30). If (6.30) does not hold we can introduce a new coordinate system say $e_1',\dots,e_n'$, with $e_n'=\xi$ in (6.26). Then calculating partial derivatives of $\bar{u},\bar{h}$ in this new coordinate system we deduce that (for $1\le i,j\le n-1$)

(6.43)
$$(\bar{h}_{X_i'X_j'}) \quad\text{is the inverse of the positive definite matrix}\quad \left(\frac{-\bar{u}_{x_i'x_j'}}{|\nabla\bar{u}|}\right).$$

Here $\bar{h}_{X_i'X_j'}$, $\bar{u}_{x_i'x_j'}$, denote second directional derivatives of $\bar{h},\bar{u}$ in the direction of $e_i'\,e_j'$ for $1\le i,j\le n-1$. Using (6.43) we can repeat the argument after (6.32) with x',X' replacing x,X to eventually conclude (6.42) holds at (ξ,t).

Hence we can continue with (6.42). Since $\xi\in\mathbb{S}^{n-1}$ and $t<t_0$ are arbitrary and $\bar{x}_1,\bar{x}_2$ are smooth we conclude for fixed t that there exist $a,b\in\mathbb{R}$ with
$$\bar{x}_1(X,t)=a\bar{x}_2(X,t)+b \quad\text{whenever}\quad X\in\mathbb{S}^{n-1}.$$

Using Remark 2.3 and the maximum principle for $\mathcal{A}$-harmonic functions we conclude that
$$u_2(x)=u_1(ax+b) \quad\text{whenever}\quad u_2(x)<t \text{ and } t<t_0.$$

To finish the proof of Theorem A let $v(x) = u_2(x) - u_1(ax+b)$ for $x \in \mathbb{R}^n$. Fix $s \in (0,1)$ and let

$$F_1 = \{x : u_1(ax+b) \geq s\} \quad \text{and} \quad F_2 = \{x : u_2(x) \geq s\}.$$

if $F_1 \neq F_2$, we assume, as we may, that z lies in the interior of $F_2 \setminus F_1$. Fix y in the interior of F_1 and draw the ray from y through z to ∞. Let w denote the point on this ray in F_2 whose distance is furtherest from y. Let l denote the part of this ray joining w to ∞. If $x \in l, x \neq w$, then since F_1, F_2 are convex and $\mathcal{A}$-harmonic functions are invariant under translation and dilation, it follows from (4.12) (a) that

$$\langle \nabla u_2(x), x-w \rangle < 0 \quad \text{and} \quad \langle \nabla u_1(ax+b), x-w \rangle < 0. \tag{6.44}$$

From arbitrariness of x and (3.2) $(\hat{a})$ it now follows that there is a connected open set, say O, containing all points in l except possibly w, for which (6.44) holds whenever $x \in O$. Clearly this inequality implies that

$$|\tau \nabla u_2(x) + (1-\tau)\nabla u_1(ax+b)| \neq 0 \quad \text{when } x \in O \text{ and } \tau \in [0,1]. \tag{6.45}$$

Using the same argument as in either (6.3)-(6.6) or (5.7)-(5.8) with $\mathcal{A} = \nabla f$ we deduce that v is a weak solution to

$$\hat{L}v = \sum_{i,j=1}^{n} \frac{\partial}{\partial x_i}(\hat{a}_{ij} v_{x_j}) = 0 \text{ in } O$$

where

$$\hat{a}_{ij}(x) = \int_0^1 \frac{\partial^2 f}{\partial \eta_i \partial \eta_j}(\tau \nabla u_2(x) + (1-\tau)\nabla u_1(ax+b))\, d\tau$$

From this deduction, Definition 2.1, (2.8), (3.3), and (6.45), we see that v is a weak solution to a locally uniformly elliptic PDE in divergence form with Lipschitz continuous and symmetric coefficients. Since $v \equiv 0$ in a neighborhood of ∞ it now follows from a unique continuation theorem (see for example [**GL**]) that v vanishes in O. Then by continuity $v(w) = 0$ so w is also in F_1. We have reached a contradiction. Thus $F_1 = F_2$ whenever $s > 0$ and consequently $v \equiv 0$ on $\mathbb{R}^n$.

CHAPTER 7

Appendix

7.1. Construction of a barrier in (4.17)

In this chapter we construct a barrier to justify display (4.17) for $1-u$. Recall that u is the $\mathcal{A}$-harmonic capacitary function in Lemma 4.3. Let $\hat{w}, \delta$ be as in (4.16) and put $\tilde{\mathcal{A}}(\eta) = -\mathcal{A}(-\eta)$ whenever $\eta \in \mathbb{R}^n$. Let $\epsilon > 0$ be given and small. We define

$$\tilde{\mathcal{A}}(\eta, \epsilon) := \int_{\mathbb{R}^n} \tilde{\mathcal{A}}(\eta - x)\theta_\epsilon(x)dx$$

whenever $\eta \in \mathbb{R}^n$ and $\theta \in C_0^\infty(B(0,1))$ with

$$\int_{\mathbb{R}^n} \theta(x)dx = 1 \text{ and } \theta_\epsilon(x) = \epsilon^{-n}\theta(x/\epsilon) \text{ for } x \in \mathbb{R}^n.$$

From Definition 2.1 and well-known properties of approximations to the identity, it follows that there exists $c = c(p,n) \geq 1$ such that

$$(c\alpha)^{-1}(\epsilon + |\eta|)^{p-2}|\xi|^2 \leq \sum_{i,j=1}^n \frac{\partial \tilde{\mathcal{A}}_i}{\partial \eta_j}(\eta, \epsilon)\xi_i\xi_j \leq c\alpha(\epsilon + |\eta|)^{p-2}|\xi|^2. \tag{7.1}$$

Note that $\tilde{\mathcal{A}}(\cdot, \epsilon)$ is infinitely differentiable for fixed $\epsilon > 0$. Let $v(\cdot, \epsilon)$ be the solution to

$$\nabla \cdot \tilde{\mathcal{A}}(\nabla v(z, \epsilon), \epsilon) = 0$$

with continuous boundary values equal to $1-u$ on $\partial B(\hat{w}, 1)$. Let

$$\tilde{\mathcal{A}}^*_{ij}(z, \epsilon) = \frac{1}{2}(\epsilon + |\nabla v(z, \epsilon)|)^{2-p}\left[\frac{\partial \tilde{\mathcal{A}}_i}{\partial \eta_j}(\nabla v(z, \epsilon), \epsilon) + \frac{\partial \tilde{\mathcal{A}}_j}{\partial \eta_i}(\nabla v(z, \epsilon), \epsilon)\right]$$

whenever $z \in B(\hat{w}, 1)$ and $1 \leq i, j \leq n$. Note also that the ellipticity constant for $\{\tilde{\mathcal{A}}^*_{ij}(z, \epsilon)\}$ and the L^∞-norm for $\tilde{\mathcal{A}}^*_{ij}$, $1 \leq i, j \leq n$, in $B(\hat{w}, 1)$ depend only on α, p, n. From (7.1) and Schauder type estimates we see that $v(\cdot, \epsilon)$ is a classical solution to the non-divergence form uniformly elliptic equation,

$$\mathcal{L}^* v = \sum_{i,j=1}^n \tilde{\mathcal{A}}^*_{ij}(z, \epsilon) v_{y_i y_j}(z) = 0$$

for $z \in B(\hat{w}, 1)$. Moreover, if we let

$$\psi(z) = \frac{e^{-N|z-\hat{w}|^2} - e^{-N}}{e^{-N/4} - e^{-N}}$$

whenever $z \in B(\hat{w}, 1) \setminus \bar{B}(\hat{w}, 1/2)$. Then $\mathcal{L}^*\psi \geq 0$ in $B(\hat{w}, 1) \setminus \bar{B}(\hat{w}, 1/2)$ if $N = N(\alpha, p, n)$ is sufficiently large, so ψ is a subsolution to $\mathcal{L}^*$ in $B(\hat{w}, 1) \setminus \bar{B}(\hat{w}, 1/2)$. Also by construction of ψ, we have $\psi = 1$ on $\partial B(\hat{w}, 1/2)$ and $\psi = 0$ on $\partial B(\hat{w}, 1)$.

Comparing boundary values of $v(\cdot,\epsilon),\psi$ and using the maximum principle for $\mathcal{L}^*$ we conclude that

$$v \geq (\min_{\bar{B}(\hat{w},1/2)} v)\,\psi \quad \text{in } B(\hat{w},1)\setminus\bar{B}(\hat{w},1/2).$$

Moreover, it is easily checked that for some $c=c(p,n,\alpha)\geq 1$

$$c\,\psi(z)\geq(1-|\hat{w}-z|) \quad \text{whenever } z\in B(\hat{w},1)\setminus\bar{B}(\hat{w},1/2).$$

Thus

$$\hat{c}\,v(z,\epsilon)) \geq (1-|\hat{w}-z|)\min_{\bar{B}(\hat{w},1/2)} v \quad \text{whenever } z\in B(\hat{w},1)\setminus\bar{B}(\hat{w},1/2) \tag{7.2}$$

for some $\hat{c}=\hat{c}(p,n,\alpha)\geq 1$. We note from Lemmas 3.1, 3.2, that a subsequence of $\{1-v(\cdot,\epsilon)\}$ converges uniformly on compact subsets of $B(\hat{w},1)$ to an $\mathcal{A}$-harmonic function in $B(\hat{w},1)$. Also by the same reasoning as in the proof of Lemma 3.3 (ii) one can derive Hölder continuity estimates for v near $\partial B(\hat{w},1)$ which are independent of ϵ. Using these facts and letting $\epsilon\to 0$ we see that a subsequence of $\{v(\cdot,\epsilon)\}$ converges uniformly on $\bar{B}(\hat{w},1)$ to $1-u$. In view of (7.2) and (4.16) we have

$$c(1-u(z))\geq\delta\,(1-|\hat{w}-z|) = \delta\,d(z,\partial B(\hat{w},1))$$

whenever $z\in B(\hat{w},1)\setminus\bar{B}(\hat{w},1/2)$ which is (4.17).

7.2. Curvature estimates for the levels of fundamental solutions

In this section we prove Lemma 6.2 when $\mathcal{A}\in M_p(\alpha)$ can be written in the form (see (2.8)):

$$\mathcal{A}_i=\frac{\partial f}{\partial\eta_i}(\eta),\ 1\leq i\leq n, \text{ where } f(t\eta)=t^p f(\eta) \text{ when } t>0,\ \eta\in\mathbb{R}^n\setminus\{0\} \tag{7.3}$$

and f has continuous second partials on $\mathbb{R}^n\setminus\{0\}$. The proof is based on some ideas garnered from reading [**CS1**]. To begin we write $f(\eta)=(k(-\eta))^p$ and note from (7.3) that $k(\eta)$ for $\eta\in\mathbb{R}^n\setminus\{0\}$ is homogeneous of degree 1 and has continuous second partials on $\mathbb{R}^n\setminus\{0\}$. We claim that

$$k^2 \text{ is strictly convex on } \mathbb{R}^n. \tag{7.4}$$

To prove (7.4) let $\lambda\in\{\eta: k(\eta)=1\}$ and put

$$\Lambda=\{\xi\in\mathbb{S}^{n-1}:\langle\nabla k(\lambda),\xi\rangle=0\}.$$

From convexity of f on $\mathbb{R}^n$ (see Definition 2.1 (i)) and the definition of k we see first that

$$f_{\eta_i\eta_j}(-\eta)=p(p-1)\,k_{\eta_i}(\eta)\,k_{\eta_j}(\eta)k^{p-2}+p\,k^{p-1}(\eta)\,k_{\eta_i\eta_j}(\eta)$$

for $1\leq i,j\leq n$. Thereupon we conclude for some $c\geq 1$ depending only on the data that if $\xi\in\Lambda$ then

$$c^{-1}\leq f_{\xi\xi}(-\lambda)=p\,k_{\xi\xi}(\lambda)\leq c. \tag{7.5}$$

Next we observe from 1-homogeneity of k that λ is an eigenvector corresponding to the eigenvalue 0 for the Hessian of k evaluated at λ. Also

$$\langle\nabla k(\lambda),\lambda\rangle=k(\lambda)\approx 1$$

so we can write

$$\tau=\nabla k(\lambda)/|\nabla k(\lambda)|=a\lambda+b\,\xi$$

where $\xi \in \Lambda$ and $a \approx 1$. Again all ratio constants depend only on the data. We conclude from (7.5) and the above facts that

$$k_{\tau\tau} \geq b^2\, k_{\xi\xi} \geq 0.$$

Thus k is positive semidefinite and an easy calculation using the above facts now gives (7.4).

From (7.4) we see as in (6.23) that if $X \in \mathbb{R}^n \setminus \{0\}$, then

$$h(X) = \sup\{\langle \eta, X\rangle :\ \eta \in \{k \leq 1\}\}$$

has continuous second partials and h is homogeneous of degree 1. Moreover,

$$\nabla h(X) = \eta(X) \text{ where } \eta \text{ is the point in } \{k=1\} \text{ with } \frac{X}{|X|} = \frac{\nabla k(\eta)}{|\nabla k(\eta)|}. \tag{7.6}$$

From calculus and Euler's formula for 1-homogeneous functions it now follows that if $X \in \mathbb{S}^{n-1}$ then

$$h(X) = \langle \eta(X), X\rangle = |X| \langle\, \eta(X), \frac{\nabla k(\eta)}{|\nabla k(\eta)|}\,\rangle = \frac{|X|}{|\nabla k(\eta)|}.$$

Using this equality we obtain first

$$\nabla k(\nabla h(X)) = \frac{|\nabla k(\eta)| X}{|X|} = \frac{X}{h(X)}$$

and second using 1-homogeneity of k, h as well as 0-homogeneity of $\nabla k, \nabla h$, that

$$k[h(X)\,\nabla h(X)]\ \nabla k[h(X)\nabla h(X)] = h(X)\, k[\nabla h(X)](X/h(X)) = X. \tag{7.7}$$

Thus $k\,\nabla k$ and $h\,\nabla h$ are inverses of each other on $\mathbb{R}^n \setminus \{0\}$.

For fixed p, $1 < p < n$, let $\beta = (p-n)/(p-1) < 0$ and define

$$\hat{G}(X) = h(X)^{\beta} \quad \text{whenever } X \in \mathbb{R}^n \setminus \{0\}.$$

We claim that $\hat{G}$ is a constant multiple of the fundamental solution for the $\mathcal{A}$ in (7.3). Indeed, if $X \in \mathbb{R}^n \setminus \{0\}$, it follows from (7.6)-(7.7) that

$$\begin{aligned}(\nabla f)(\nabla \hat{G}(X)) &= -p\, k^{p-1}(-\nabla \hat{G}(X))\ (\nabla k(-\nabla \hat{G}(X)))\\ &= \frac{-X}{h(X)}\, p\, k^{p-1}(-\nabla \hat{G}(x))\\ &= Xp(-\beta)^{p-1}\, h^{[(\beta-1)(p-1)-1]}(X)\, k^{p-1}(\nabla h(X))\\ &= X\, p\, (-\beta)^{p-1} h(X)^{-n}.\end{aligned} \tag{7.8}$$

Now $X \mapsto h(X/|X|)^{-n}$ is homogeneous of degree 0 so

$$\langle X, \nabla[h(X/|X|)^{-n}]\rangle = 0$$

by Euler's formula. From this observation and (7.8) we deduce

$$\begin{aligned}&p^{-1}(-\beta)^{1-p}\nabla\cdot\Big((\nabla f)(\nabla \hat{G}(X))\Big)\\ &\qquad = h(X/|X|)^{-n}\ \nabla\cdot(X|X|^{-n}) + |X|^{-n}\langle X, \nabla[h(X/|X|)^{-n}]\rangle\\ &\qquad = 0\end{aligned}$$

when $X \in \mathbb{R}^n \setminus \{0\}$. Hence $\hat{G}$ is $\mathcal{A}$-harmonic in $\mathbb{R}^n \setminus \{0\}$. Now from 1-homogeneity of h and (7.6) it is easily seen that (5.1) $(a), (b)$, are valid for $\hat{G}$ with constants that depend only on p, n, α. Also from (7.8) we note that

$$|\nabla f(\nabla \hat{G}(X))| \approx |X|^{1-n} \text{ on } \mathbb{R}^n \setminus \{0\}.$$

If $\theta \in C_0^\infty(\mathbb{R}^n)$ then from the above display we deduce that the function $X \mapsto \langle \nabla f(\nabla \hat{G}(X)), \nabla\theta(X)\rangle$ is integrable on $\mathbb{R}^n$. Using this fact, smoothness of f, h, and an integration by parts, we get

(7.9)

$$\int_{\mathbb{R}^n} \langle \nabla f(\nabla \hat{G}(X)), \nabla\theta(X)\rangle dx = -\lim_{r\to 0}\int_{\partial B(0,r)} \theta(X)\,\langle \nabla f(\nabla \hat{G}(X)), \frac{X}{|X|}\rangle\, d\mathcal{H}^{n-1}$$
$$= b\,\theta(0).$$

Using (7.8) once again it follows that

$$b = -\lim_{r\to 0}\int_{\partial B(0,r)} \langle \nabla f(\nabla \hat{G}(X)), X/|X|\rangle\, d\mathcal{H}^{n-1}$$
$$= p(-\beta)^{p-1}\int_{\partial B(0,1)} h(X/|X|)^{-n}\, d\mathcal{H}^{n-1}.$$

From (7.9) and (5.1) (a), (b), we conclude from (5.1) (d) that $\hat{G}$ is a constant multiple of the fundamental solution for $\mathcal{A} = \nabla f$.

To prove Lemma 6.2 which says that (6.1) holds for G, we show that $\hat{G}$ satisfies (6.1) which will finish the proof. To this end, recall that $k\nabla k$ and $h\nabla h$ are inverse functions. Thus, by the chain rule the $n \times n$ matrices

(7.10) $\quad (k\, k_{\eta_i\eta_j} + k_{\eta_i}k_{\eta_j})$ and $(h\, h_{X_iX_j} + h_{X_i}h_{X_j})$ are inverses of each other.

From (7.4) and (7.10) we conclude that $(h\, h_{\eta_i\eta_j} + h_{\eta_i}h_{\eta_j})$ is homogeneous of degree 0 and positive definite with eigenvalues bounded above and below by constants depending only on p, n, α.

To prove (6.1) for $\hat{G}$, suppose $X, \xi \in \mathbb{S}^{n-1}$ and $\langle \nabla \hat{G}(X), \xi\rangle = 0$. As $\hat{G} = h^\beta$ (where $\beta < 0$) we also have $\langle \nabla h(X), \xi\rangle = 0$ and

$$-\hat{G}_{\xi\xi} = -\beta h^{\beta-1}\, h_{\xi\xi}(X) \geq \tau' > 0.$$

From this inequality and (5.1) (b) or (7.6) we see that τ' depends only on the data. Thus (6.1) holds and proof of Lemma 6.2 is complete.

REMARK 7.1. *In view of* (7.9) *and* (7.8)

$$G(x) = b^{\frac{-1}{p-1}}\,\hat{G}(x) = b^{\frac{-1}{p-1}}\,h(x)^{\frac{p-n}{p-1}}$$

is the fundamental solution in Lemma 5.1 *where*

$$b = c\int_{\partial B(0,1)} h(X/|X|)^{-n}\, d\mathcal{H}^{n-1} = p(-\beta)^{p-1}\int_{\partial B(0,1)} h(X/|X|)^{-n}\, d\mathcal{H}^{n-1}$$
$$= p\left(\frac{n-p}{p-1}\right)^{p-1}\int_{\mathbb{S}^{n-1}} h(\omega)^{-n}\, d\omega.$$

Part 2

A Minkowski problem for nonlinear capacity

CHAPTER 8

Introduction and statement of results

In this chapter we use our work on the Brunn-Minkowski inequality to study the Minkowski problem associated with $\mathcal{A} = \nabla f$-capacities when f is as in Theorem A. To be more specific, suppose $E \subset \mathbb{R}^n$ is a compact convex set with nonempty interior. Then for $\mathcal{H}^{n-1}$ almost every $x \in \partial E$, there is a well defined outer unit normal, $\mathbf{g}(x)$ to ∂E. The function $\mathbf{g} : \partial E \to \mathbb{S}^{n-1}$ (whenever defined), is called the Gauss map for ∂E. The problem originally considered by Minkowski states: given a positive finite Borel measure μ on $\mathbb{S}^{n-1}$ satisfying

$$
\begin{aligned}
&(i) \int_{\mathbb{S}^{n-1}} |\langle \theta, \zeta \rangle| \, d\mu(\zeta) > 0 \text{ for all } \theta \in \mathbb{S}^{n-1}, \\
&(ii) \int_{\mathbb{S}^{n-1}} \zeta \, d\mu(\zeta) = 0,
\end{aligned}
\tag{8.1}
$$

show there exists up to translation a unique compact convex set E with nonempty interior and

$$\mathcal{H}^{n-1}(\mathbf{g}^{-1}(K)) = \mu(K) \text{ whenever } K \subset \mathbb{S}^{n-1} \text{ is a Borel set.}$$

Minkowski [**M1, M2**] proved existence and uniqueness of E when μ is discrete or has a continuous density. The general case was treated by Alexandrov in [**A1, A2**] and Fenchel and Jessen in [**FJ**]. Note also that the conditions in (8.1) are also necessary conditions for the existence and uniqueness of measure μ.

In [**J**], a similar problem was considered for electrostatic capacity when $E \subset \mathbb{R}^n, n \geq 3$, is a compact convex set with nonempty interior and u is the Newtonian or 2-capacitary function for E. In this case, u is harmonic in $\Omega = \mathbb{R}^n \setminus E$ with boundary value 1 on ∂E and goes to zero as $|x| \to \infty$. Then a well-known work of Dahlberg [**D**] implies that

$$\lim_{\substack{y \to x \\ y \in \Gamma(x)}} \nabla u(y) = \nabla u(x) \text{ exists for } \mathcal{H}^{n-1} \text{ almost every } x \in E. \tag{8.2}$$

Here $\Gamma(x)$ is the non-tangential approach region in $\mathbb{R}^n \setminus E$. Also,

$$\int_{\partial E} |\nabla u(x)|^2 d\mathcal{H}^{n-1} < \infty.$$

If μ is a positive finite Borel measure on $\mathbb{S}^{n-1}$ satisfying (8.1), it is shown by Jerison in [**J**, Theorem 0.8] that there exists E a compact convex set with nonempty interior and corresponding 2-capacity function u with

$$\int_{\mathbf{g}^{-1}(K)} |\nabla u(x)|^2 \, d\mathcal{H}^{n-1} = \mu(K) \tag{8.3}$$

whenever $K \subset \mathbb{S}^{n-1}$ is a Borel set and $n \geq 4$. Moreover, E is the unique compact convex set with nonempty interior up to translation for which (8.3) holds. If $n = 3$, a less precise result is available.

Jerison's result was generalized in [**CNSXYZ**] as follows. Given a compact convex set E with nonempty interior and p fixed, $1 < p < n$, let u be the p-capacitary function for E. Then from [**LN**, Theorem 3] it follows that (8.2) holds for u. Thus the Gauss map $\mathbf{g}$ can be defined for $\mathcal{H}^{n-1}$-almost every $x \in \partial E$. If μ is a positive finite Borel measure on $\mathbb{S}^{n-1}$ having no antipodal point masses (i.e., it is not true that $0 < \mu(\{\xi\}) = \mu(\{-\xi\})$ for some $\xi \in \mathbb{S}^{n-1}$) and if (8.1) holds, then it is shown in [**CNSXYZ**] that for $1 < p < 2$, there exists E a compact convex set with nonempty interior and corresponding p-capacitary function u with

$$\int_{\mathbf{g}^{-1}(K)} |\nabla u(x)|^p \, d\mathcal{H}^{n-1} = \mu(K) \tag{8.4}$$

whenever $K \subset \mathbb{S}^{n-1}$ is a Borel set. Assuming the existence of an E for which (8.4) holds when p is fixed, $1 < p < n$, it was also shown in [**CNSXYZ**] that E is unique up to translation when $p \neq n-1$ and unique up to translation and dilation when $p = n-1$.

We consider an analogous problem:

THEOREM B. *Let μ is a positive finite Borel measure on $\mathbb{S}^{n-1}$ satisfying* (8.1). *Let p be fixed, $1 < p < n$ and $\mathcal{A} = \nabla f$ as in* (2.7) *in Theorem* A.

(8.5)

If $p \neq n-1$ then there exists a compact convex set E with nonempty interior and corresponding $\mathcal{A}$-capacitary function u satisfying

(*a*) (8.2) *holds for u and* $\displaystyle\int_{\partial E} f(\nabla u(x)) \, d\mathcal{H}^{n-1} < \infty.$

(*b*) $\displaystyle\int_{\mathbf{g}^{-1}(K)} f(\nabla u(x)) \, d\mathcal{H}^{n-1} = \mu(K)$ *whenever $K \subset \mathbb{S}^{n-1}$ is a Borel set.*

(*c*) *E is the unique set up to translation for which* (*b*) *holds.*

If $p = n-1$ then there exists a compact convex set E with nonempty interior, a constant $b \in (0, \infty)$, and corresponding $\mathcal{A}$-capacitary function u satisfying (*a*) *and*

(*d*) $\displaystyle b \int_{\mathbf{g}^{-1}(K)} f(\nabla u) \, d\mathcal{H}^{n-1} = \mu(K)$ *whenever $K \subset \mathbb{S}^{n-1}$ is a Borel set.*

(*e*) *E is the unique set up to translation satisfying* (*d*) *and* $\operatorname{Cap}_{\mathcal{A}}(E) = 1$.

As a broad outline of our proof we follow [**CNSXYZ**] (who in turn used ideas from [**J**]). However, several important arguments in [**CNSXYZ**] used tools from [**LN, LN1, LN2**] for p-harmonic functions vanishing on a portion of the boundary of a Lipschitz domain. To our knowledge similar results have not yet been proved for $\mathcal{A} = \nabla f$-harmonic functions and the arguments although often straight forward for the experts are rather subtle. In reviewing these arguments we naturally made editing decisions as to which details to include and which to refer to. Also we attempted to clarify some details that were not obvious to us even in the p-harmonic case and our proofs sometimes use later work of the fourth named author and Nyström in [**LN3, LN4**] when the authors "could see the forest for the trees". Thus the reader is advised to have the above papers on hand. These preliminary results for the proof of Theorem B are given in chapters 9 and 10. Our work in these chapters gives (a) in Theorem B. In chapter 11 we consider a sequence of compact convex sets, say $\{E_m\}_{m\geq 1}$ with nonempty interiors which converge in the sense of Hausdorff distance to E a compact convex set. If $\{\mu_m\}_{m\geq 1}$ and μ denote the corresponding measures as in (8.5) we show that $\{\mu_m\}$ converges weakly to μ on $\mathbb{S}^{n-1}$. In chapter 12 we first derive the Hadamard variational formula for $\mathcal{A} = \nabla f$-capacitary functions in compact convex sets with nonempty interior and smooth boundary. Second using the results in chapter 11 and taking limits we get this formula for an arbitrary compact convex set with nonempty interior. Finally, in chapter 13 we consider a minimum problem similar to the one considered in [**J**, **CNSXYZ**]. However, unlike [**CNSXYZ**], we are able to show that compact convex sets of dimension $k \leq n-1$ (so with empty interior) cannot be a solution to our minimum problem. To rule out these possibilities we use work in [**LN4**] when $k < n-1$ while if $k = n-1$ we use an argument of Venouziou and Verchota in [**VV**]. The solution to this minimum problem gives existence of E in Theorem B while uniqueness is proved using Theorem A.

CHAPTER 9

Boundary behavior of $\mathcal{A}$-harmonic functions in Lipschitz domains

We begin this chapter with several definitions. Recall that $\phi : K \to \mathbb{R}$ is said to be Lipschitz on K provided there exists $\hat{b}, 0 < \hat{b} < \infty$, such that

$$|\phi(z) - \phi(w)| \leq \hat{b}\,|z - w| \quad \text{whenever } z, w \in K. \tag{9.1}$$

The infimum of all $\hat{b}$ such that (9.1) holds is called the Lipschitz norm of ϕ on K, denoted $\|\phi\hat{\|}_K$. It is well-known that if $K = \mathbb{R}^{n-1}$, then ϕ is differentiable almost everywhere on $\mathbb{R}^{n-1}$ and $\|\phi\hat{\|}_{\mathbb{R}^{n-1}} = \|\,|\nabla\phi|\,\|_\infty$.

DEFINITION 9.1 (**Lipschitz Domain**). *A domain $D \subset \mathbb{R}^n$ is called a bounded Lipschitz domain provided that there exists a finite set of balls $\{B(x_i, r_i)\}$ with $x_i \in \partial D$ and $r_i > 0$, such that $\{B(x_i, r_i)\}$ constitutes a covering of an open neighborhood of ∂D and such that, for each i,*

$$D \cap B(x_i, 4r_i) = \{y = (y', y_n) \in \mathbb{R}^n : y_n > \phi_i(y')\} \cap B(x_i, 4r_i),$$
$$\partial D \cap B(x_i, 4r_i) = \{y = (y', y_n) \in \mathbb{R}^n : y_n = \phi_i(y')\} \cap B(x_i, 4r_i),$$

in an appropriate coordinate system and for a Lipschitz function ϕ_i on $\mathbb{R}^{n-1}$. The Lipschitz constant of D is defined to be $M = \max_i \||\nabla\phi_i|\|_\infty$.

If D is Lipschitz and $r_0 = \min r_i$, then for each $w \in \partial D$, $0 < r < r_0$, we can find points

$$a_r(w) \in D \cap B(w, r) \quad \text{with} \quad d(a_r(w), \partial D) \geq c^{-1} r$$

for a constant $c = c(M)$. In the following, we let $a_r(w)$ denote one such point. We also put $\Delta(w, r) = \partial D \cap B(w, r)$ when $w \in \partial D$ and $r > 0$.

DEFINITION 9.2 (**Starlike Lipschitz domain**). *A bounded domain $D \subset \mathbb{R}^n$ is said to be starlike Lipschitz with respect to $z \in D$ provided*

$$\partial D = \{z + \mathcal{R}(\omega)\omega : \omega \in \partial B(0,1)\}$$
$$\qquad \text{where } \log \mathcal{R} : \partial B(0,1) \to \mathbb{R} \text{ is Lipschitz on } \partial B(0,1).$$

Under the above scenario we say that z is the center of D and $\|\log \mathcal{R}\hat{\|}_{\mathbb{S}^{n-1}}$ is the starlike Lipschitz constant for D. In the rest of this chapter reference to the "data" means the constants in Definition 2.1, (4.8) for $\mathcal{A} = \nabla f$, p, n, and the Lipschitz or starlike Lipschitz constant whenever applicable. We shall need some lemmas similar to Lemmas 3.3, 3.4 for $\mathcal{A} = \nabla f$-harmonic functions vanishing on a portion of a Lipschitz or starlike Lipschitz domain. In the next two lemmas, $r_0' = r_0$ when D is Lipschitz and $r_0' = |w - z|/100$ when D is starlike Lipschitz with center at z.

LEMMA 9.3. *Let $D \subset \mathbb{R}^n$ be a bounded Lipschitz or starlike Lipschitz domain with center at z and p fixed, $1 < p < n$. Let $w \in \partial D$, $0 < 4r < r_0'$, and suppose that v is a positive $\mathcal{A}$-harmonic function in $D \cap B(w, 4r)$ with $v \equiv 0$ on $\partial D \cap B(w, 4r)$ in the $W^{1,p}$ Sobolev sense. Then v has a representative in $W^{1,p}(D \cap B(w, s))$, $s < 4r$, which extends to a Hölder continuous function on $B(w, s)$ (denoted v) with $v \equiv 0$ on $B(w, s) \setminus D$. Also, there exists $c \geq 1$, depending only on the data, such that if $\bar{r} = r/c$, then*

$$\bar{r}^{p-n} \int_{B(w,\bar{r})} |\nabla v|^p dx \leq c(v(a_{2\bar{r}}(w)))^p. \tag{9.2}$$

Moreover, there exists $\tilde{\beta} \in (0, 1)$, depending only on the data, such that if $x, y \in B(w, \bar{r})$, then

$$|v(x) - v(y)| \leq c \left(\frac{|x - y|}{\bar{r}} \right)^{\tilde{\beta}} v(a_{\bar{r}}(w)). \tag{9.3}$$

PROOF. Here (9.2) with $v(a_{2\bar{r}}(w))$ replaced by $\max_{B(w,2\bar{r})} v$ is just a standard Caccioppoli inequality while (9.3) with $v(a_{2\bar{r}}(w))$ replaced by $\max_{B(w,2\bar{r})} v$ follows as in Lemma 3.3 from (3.5) with E replaced by $\Delta(w, \bar{r})$ and Theorem 6.18 in [**HKM**]. The fact that $\max_{B(w,2\bar{r})} v \approx v(a_{2\bar{r}}(w))$ follows from an argument often attributed to several authors (see in [**LN**, Lemma 2.2]). □

In the sequel, we always assume v as above $\equiv 0$ on $B(w, 4r) \setminus D$.

LEMMA 9.4. *Let D, v, p, r, w be as in Lemma 9.3. There exists a unique finite positive Borel measure τ on $\mathbb{R}^n$, with support contained in $\bar{\Delta}(w, r)$, such that if $\phi \in C_0^\infty(B(w, r))$ then*

$$\int \langle \nabla f(\nabla v), \nabla \phi \rangle dx = - \int \phi \, d\tau. \tag{9.4}$$

Moreover, there exists $c \geq 1$ depending only on the data such that if $\bar{r} = r/c$, then

$$c^{-1} \bar{r}^{p-n} \tau(\Delta(w, \bar{r})) \leq (v(a_{\bar{r}}(w)))^{p-1} \leq c \bar{r}^{p-n} \tau(\Delta(w, \bar{r})). \tag{9.5}$$

PROOF. See [**KZ**, Lemma 3.1] for a proof of Lemma 9.4. □

We note that lemmas similar to Lemmas 9.5-9.6 and Proposition 9.7 which follow are proved for p-harmonic functions in [**LN**, Lemmas 2.5, 2.39, 2.45].

Throughout the remainder of this paper, we assume that $\mathcal{A} = \nabla f$ where f is as in Theorem A. In order to state the next lemma, we need some more notation. Let D be a starlike Lipschitz domain with center at z. Given $x \in \partial D$ and $b > 1$, let

$$\Gamma(x) = \{y \in D : |y - x| < b \, d(y, \partial D)\}.$$

If $w \in \partial D$, $0 < r \leq |w - z|/100$, and $x \in \partial D \cap B(w, r)$, we note from elementary geometry that if b is large enough (depending on the starlike Lipschitz constant for D) then $\Gamma(x) \cap B(w, 8r)$ contains the inside of a truncated cone with vertex x, axis parallel to $z - x$, angle opening $\theta = \theta(b) > 0$, and height r. Fix b so that this property holds for all $x \in \partial D$. Given a measurable function g on $D \cap B(w, 8r)$ define the *non-tangential maximal function*

$$\mathcal{N}_r(g) : \partial D \cap B(w, r) \to \mathbb{R}$$

of g relative to $D \cap B(w,r)$ by

$$\mathcal{N}_r(g)(x) = \sup_{y \in \Gamma(x) \cap B(w,8r)} |g|(y) \quad \text{whenever } x \in \partial D \cap B(w,r).$$

Next we prove a reverse Hölder inequality.

LEMMA 9.5 (**Reverse Hölder inequality**). *Let D be a starlike Lipschitz domain with center z and let $w \in \partial D$ with $0 < r < |w-z|/100$. Let v, τ, be as in Lemma 9.4 and suppose that for some $c_\star \geq 1$,*

$$(9.6) \qquad c_\star^{-1} \frac{v(x)}{d(x,\partial D)} \leq \langle \frac{z-x}{|z-x|}, \nabla v(x) \rangle \leq |\nabla v(x)| \leq c_\star \frac{v(x)}{d(x,\partial D)}$$

whenever $x \in B(w,4r) \cap D$. There exists $c \geq 1$, depending only on $c_\star$ and the data, such that if $\tilde r = r/c$, then

$$\frac{d\tau}{d\mathcal{H}^{n-1}}(y) = k^{p-1}(y) \quad \text{for } y \in \Delta(w,\tilde r).$$

Also, there exists $q > p, c_1$, and c_2 depending only on $c_\star$ and the data with

(9.7)

$$(a) \quad \int_{\Delta(w,\tilde r)} k^q \, d\mathcal{H}^{n-1} \leq c_1 \, \tilde r^{\frac{(n-1)(p-1-q)}{p-1}} \left(\int_{\Delta(w,\tilde r)} k^{p-1} \, d\mathcal{H}^{n-1} \right)^{q/(p-1)}.$$

$$(b) \quad \int_{\Delta(w,\tilde r)} \mathcal{N}_{\tilde r}(|\nabla v|)^q \, d\mathcal{H}^{n-1} \leq c_2 \, \tilde r^{\frac{(n-1)(p-1-q)}{p-1}} \left(\int_{\Delta(w,\tilde r)} k^{p-1} \, d\mathcal{H}^{n-1} \right)^{q/(p-1)}.$$

PROOF. Let $\tilde r = r/c$ where $c \geq 100$ is to be determined and for fixed $s, \tilde r < s < 2\tilde r$ and $t > 0$ small, let

$$D_1 = B(w,s) \cap D \cap \{v > t\}.$$

Since $\mathcal{A}$-harmonic functions are invariant under translation, we assume as we may that $z = 0$. Note from (3.3) that $\partial D_1 \cap B(w,s)$ is smooth with outer normal $\nu = -\nabla v/|\nabla v|$ and also that we can apply the divergence theorem to $xf(\nabla v(x))$ in D_1. Doing this and using $\mathcal{A}$-harmonicity of v in D_1, we arrive at

$$(9.8) \qquad I = \int_{D_1} \nabla \cdot (xf(\nabla v)) \, dx = \int_{\partial D_1} \langle x, \nu \rangle f(\nabla v) d\mathcal{H}^{n-1}$$

and

$$(9.9) \qquad \begin{aligned} I &= n \int_{D_1} f(\nabla v) dx + \sum_{k,j=1}^{n} \int_{D_1} x_k f_{\eta_j}(\nabla v) v_{x_j x_k} dx \\ &= n \int_{D_1} f(\nabla v) dx + I_1. \end{aligned}$$

Integrating I_1 by parts, using p-homogeneity of f, as well as $\mathcal{A} = \nabla f$-harmonicity of v in D_1 we deduce that

$$(9.10) \qquad I_1 = \int_{\partial D_1} \langle x, \nabla v \rangle \, \langle \nabla f(\nabla v), \nu \rangle \, d\mathcal{H}^{n-1} - p \int_{D_1} f(\nabla v) dx.$$

Combining (9.8)-(9.10) we find after some juggling that

$$(n-p)\int_{D_1} f(\nabla v)dx = \int_{\partial D_1}\langle x,\nu\rangle f(\nabla v)d\mathcal{H}^{n-1} - \int_{\partial D_1}\langle x,\nabla v\rangle\,\langle\nabla f(\nabla v),\nu\rangle\, d\mathcal{H}^{n-1}. \tag{9.11}$$

From p-homogeneity of f we obtain

$$\begin{aligned}\int_{\partial D_1\cap B(w,s)}\langle x,\nu\rangle f(\nabla v)d\mathcal{H}^{n-1} - \int_{\partial D_1\cap B(w,s)}\langle x,\nabla v\rangle\,\langle\nabla f(\nabla v),\nu\rangle\, d\mathcal{H}^{n-1} \\ = (p-1)\int_{\partial D_1\cap B(w,s)}\langle x,\nabla v\rangle\,\frac{f(\nabla v)}{|\nabla v|}d\mathcal{H}^{n-1} \\ \leq 0.\end{aligned} \tag{9.12}$$

Using (9.12) in (9.11) and (9.9), (9.6), we arrive after some more juggling at

$$\begin{aligned} c^{-1}\int_{\partial D_1\cap B(w,s)}|x|f(\nabla v)d\mathcal{H}^{n-1} &\leq -(p-1)\int_{\partial D_1\cap B(w,s)}\langle x,\nabla v\rangle\frac{f(\nabla v)}{|\nabla v|}d\mathcal{H}^{n-1} \\ &\leq F_1\end{aligned} \tag{9.13}$$

where

$$F_1 = c\int_{\partial D_1\cap\partial B(w,s)}|x|f(\nabla v)d\mathcal{H}^{n-1}. \tag{9.14}$$

Here c depends only on $c_\star$ and the data. Also in getting F_1 we have used the structure assumptions on f in Theorem A.

We note from (9.3) that

$$\bar{D}\cap\bar{B}(w,2r)\cap\{v\geq t\}\to\bar{D}\cap\bar{B}(w,2r) \tag{9.15}$$

in Hausdorff distance as $t\to 0$. Also, from (9.6) we note that if $v(x)=t$ and $\omega = x/|x|$ then $\tilde{\mathcal{R}}(\omega)=|x|$ is well-defined. Moreover, if

$$\Theta = \{\omega\in\mathbb{S}^{n-1} : \omega = x/|x| \text{ for some } x\in\bar{B}(w,2r) \text{ with } v(x)=t\}$$

then $\log\tilde{\mathcal{R}}$ is Lipschitz on Θ with Lipschitz constant depending only on the data and $c_\star$. Using the Whitney extension theorem (see [**St**, Chapter VI, Section 1]), we can extend $\log\tilde{\mathcal{R}}$ to a Lipschitz function on $\mathbb{S}^{n-1}$ with Lipschitz constant depending only on $c_\star$ and the data.

We next let $\tilde{v}=\max(v-t,0)$ and if $t>0$ is sufficiently small then we see from (9.15) that Lemmas 9.3, 9.4 can be applied to $\tilde{v}$ in $D\cap B(w,2r)\cap\{v>t\}$. Let $\tilde{\tau}$ be the measure corresponding to $\tilde{v}$. Then from smoothness of $\tilde{v}$, the divergence theorem, and p-homogeneity of f we obtain that

$$d\tilde{\tau}(x) = p\frac{f(\nabla v(x))}{|\nabla v(x)|}\,d\mathcal{H}^{n-1} \quad \text{for } x\in B(w,2r)\cap\{v=t\}.$$

To estimate F_1 in (9.13) choose $s\in(\tilde{r},2\tilde{r})$ so that

$$\int_{\partial B(w,s)\cap\partial D_1} f(\nabla v)\,d\mathcal{H}^{n-1} \leq 2\tilde{r}^{-1}\int_{B(w,2\tilde{r})\cap D\cap\{v>t\}} f(\nabla v)\,dx.$$

This choice is possible from weak type estimates or Chebyshev's inequality. Using this inequality in (9.14) and the above lemmas for $\tilde{v}$ we obtain for $t>0$, small and

c sufficiently large in the definition of $\tilde{r}$, that

$$
\begin{aligned}
|w-z|^{-1}F_1 &\leq 4\tilde{r}^{-1}\int_{B(w,2\tilde{r})\cap D\cap\{v>t\}} f(\nabla v)\,dx \\
&\leq \tilde{c}\tilde{r}^{n-1-p}\,\tilde{v}(a_{4\tilde{r}}(w))^p \\
&\leq \tilde{c}^2\tilde{r}^{\frac{1-n}{p-1}}\,(\tilde{\tau}(B(w,2\tilde{r}))^{p/(p-1)}
\end{aligned}
\tag{9.16}
$$

where $\tilde{c}$ depends only on the data and we have also used Harnack's inequality for $\tilde{v}$ to get the last inequality. Putting (9.16) into (9.13), and using (9.2), (9.5), we find that if

$$\tilde{k}^{p-1}(y) = p\frac{f(\nabla v(y))}{|\nabla v(y)|} \quad \text{for} \quad y\in\{v=t\}$$

then for small $t>0$,

$$
\begin{aligned}
\int_{B(w,\tilde{r})\cap\{\tilde{v}=0\}} \tilde{k}^p\,d\mathcal{H}^{n-1} &\leq c\,\tilde{r}^{\frac{1-n}{p-1}}\left(\int_{B(w,2\tilde{r})\cap\{\tilde{v}=0\}} \tilde{k}^{p-1}\,d\mathcal{H}^{n-1}\right)^{p/(p-1)} \\
&\leq c^2\,\tilde{r}^{\frac{1-n}{p-1}}\left(\int_{B(w,\tilde{r})\cap\{\tilde{v}=0\}} \tilde{k}^{p-1}\,d\mathcal{H}^{n-1}\right)^{p/(p-1)}
\end{aligned}
\tag{9.17}
$$

where we have once again used Harnack's inequality for $\tilde{v}$ in Lemma 9.4 to get the last inequality.

With $\tilde{r}$ fixed we now let $t\to 0$ through a decreasing sequence $\{t_m\}$. Let $\tau_m=\tilde{\tau}$ when $t=t_m$ From (3.2) and Lemmas 9.3-9.4 we see that

$$\tau_m \text{ converges weakly to } \tau \text{ as } m\to\infty$$

where τ is the measure associated with v. Using the change of variables formula and Lemma 9.4 we can pull back each τ_m to a measure on a subset of $\mathbb{S}^{n-1}$. In view of (9.17) we see that the Radon-Nikodym derivative of each pullback measure with respect to $\mathcal{H}^{n-1}$ measure on $\mathbb{S}^{n-1}$ satisfies a $L^{p/(p-1)}$ reverse Hölder inequality on

$$\{x/|x| : x\in B(w,\tilde{r})\cap\{v=t_m\}\}.$$

Moreover, $L^{p/(p-1)}$ and Hölder constants depend only on $c_\star$ and the data. Thus, any sequence of these derivatives has a subsequence which converges weakly in $L^{p/(p-1)}$. Using these observations we deduce first that τ viewed as a measure on a subset of $\mathbb{S}^{n-1}$ has a density that is $p/(p-1)$ integrable and second that this density satisfies a $p/(p-1)$ reverse Hölder inequality. Transforming back we conclude that if k^{p-1} denotes the Radon-Nikodym derivative of τ on $\Delta(w,\tilde{r})$ with respect to $\mathcal{H}^{n-1}$ then (9.7) is valid with q replaced by p. Now r,w can obviously be replaced by y,ρ where $y\in B(w,r)\cap\partial D$ and $0<\rho<r/2$ in (9.7) with q replaced by p. Doing this we see from a now well-known theorem that the resulting reverse Hölder inequality is self-improving, i.e, holds for some $q>p$, depending only on $c_\star$ and the data (see [**CF**, Theorem IV] for a proof of the self improving property). This proves (9.7) (a).

To prove (9.7) (b) let $x\in D\cap B(w,\tilde{r}), y\in\Gamma(x)$, and suppose

$$\mathcal{N}_{\tilde{r}}(|\nabla v|)(x) \leq 2|\nabla v(y)| \leq 2\mathcal{N}_{\tilde{r}}(|\nabla v|)(x)).$$

If $\tilde r/100 \le d(y,\partial D)$, we deduce from (9.6), Lemma 9.4, and Harnack's inequality for v that for some $c' \ge 1$ depending only on σ, b, and the data,

$$\mathcal{N}_{\tilde r}(|\nabla v|)(x) \le c' \left(\tilde r^{1-n} \int_{\Delta(x,\tilde r)} k^{p-1}\, d\mathcal{H}^{n-1} \right)^{\frac{1}{p-1}}.$$

Otherwise,

$$\mathcal{N}_{\tilde r}(|\nabla v|)(x) \le 2|\nabla v(y))| \le c|x-y|^{-1} v(y) \le c^2\, \mathcal{M}_1 k(x)$$

where

$$\mathcal{M}_1 k(x) := \left(\sup_{0<t<\tilde r} t^{1-n} \int_{\Delta(x,t)} k^{p-1}\, d\mathcal{H}^{n-1} \right)^{\frac{1}{p-1}}. \tag{9.18}$$

Raising both sides of either inequality to the q-th power and integrating over $x \in \Delta(w,\tilde r)$, we deduce from the Hardy-Littlewood maximal theorem (see [**St**, Chapter 1]) that (9.7) (b) is true. This completes the proof of Lemma 9.5. □

We use Lemma 9.5 to prove the following localization lemma.

Lemma 9.6. *Let $v, D, z, c_\star, b, w, r, k$ be as in Lemma 9.5. Let $\hat w \in \Delta(w,2r), 0 < s < r$, and let $\hat w' \in D$ be the point on the ray from z through $\hat w$ with $|\hat w - \hat w'| = s/100$. There exists $c', c'' \ge 1$, depending only on $c_\star$ and the data, such that if $s' = s/c'$, then there is a starlike Lipschitz domain $\tilde D = \tilde D(\hat w, s) \subset B(\hat w, s) \cap D$ with center at $\hat w'$,*

$$\frac{\mathcal{H}^{n-1}[\partial \tilde D \cap \Delta(\hat w, s')]}{\mathcal{H}^{n-1}[\Delta(\hat w, s')]} \ge 3/4 \tag{9.19}$$

and Lipschitz constant $\le c\,(\|\log \hat{\mathcal{R}}\|_{\mathbb{S}^{n-1}} + 1)$ where $\mathcal{R}$ is the graph function for D, and c depends only on p, n. Moreover, if $x \in \tilde D$, then

$$\frac{1}{c''} \frac{v(\hat w')}{s} \le |\nabla v(x)| \le c'' \frac{v(\hat w')}{s}. \tag{9.20}$$

Proof. From starlike Lipschitzness of D and basic geometry we deduce the existence of $c' > 1$ (depending only on the data) such that if $s' = s/c'$, then for any $x \in \Delta(\hat w, s')$, the ice cream cone, say $C(x, \hat w')$ obtained by drawing rays from x to all points in $\bar B(\hat w', s')$ satisfies

$$\Gamma(x) \cap B(\hat w, s) \supset C(x, \hat w'). \tag{9.21}$$

for $b > 1$ suitably large. Put

$$\mathcal{M}_2 k(x) = \left(\inf_{0<t<s'} t^{1-n} \int_{\Delta(x,t)} k^{p-1} d\mathcal{H}^{n-1} \right)^{1/(p-1)}$$

for $x \in \Delta(\hat w, s')$. We claim there exists a compact set $\hat K \subset \Delta(\hat w, s')$ and $\hat c \ge 1$ (depending only on the data, as well as $c_\star$ in (9.6)) with

$$\hat c\, \mathcal{M}_2 k > v(a_{s'}(\hat w))/s' \text{ on } \hat K \quad \text{and} \quad \hat c\, \mathcal{H}^{n-1}(\hat K) > (s')^{n-1}. \tag{9.22}$$

To see this we temporarily allow $\hat c$ to vary. We note that if

$$\epsilon = (1/\hat c)\, v(a_{s'}(\hat w))/s',$$
$$\Phi = \{x \in \Delta(\hat w, s') : \mathcal{M}_2 k(x) \le \epsilon\}$$

then by a standard covering argument there exists $\{B(x_i, r_i)\}$ with $x_i \in \Phi$, $0 < r_i \leq s'$, $\Phi \subset \bigcup_i B(x_i, r_i)$ and $\{B(x_i, r_i/10)\}$ pairwise disjoint. Also,

$$\int_{\Delta(x_i, r_i)} k^{p-1}\, d\mathcal{H}^{n-1} \leq (2\epsilon)^{p-1}\, r_i^{n-1} \quad \text{for each } i.$$

Using these facts and $\mathcal{H}^{n-1}(B(x_i, r_i/10) \cap \partial D) \approx r_i^{n-1}$, we get

$$\begin{aligned} \int_{\Phi} k^{p-1}\, d\mathcal{H}^{n-1} &\leq \sum_i \int_{\Delta(x_i, r_i)} k^{p-1}\, d\mathcal{H}^{n-1} \\ &\leq (2\epsilon)^{p-1} \sum_i r_i^{n-1} \\ &\leq c\, \epsilon^{p-1}\, (s')^{n-1}. \end{aligned} \tag{9.23}$$

On the other hand, if $\Psi = \Delta(\hat{w}, s') \setminus \Phi$, then from (9.5), (9.6), and (9.7) with r replaced by s', the structure assumptions on $\mathcal{A}$, and Hölder's inequality, we get for some c, depending only on the data and $c_\star$ in (9.6),

$$\begin{aligned} \int_{\Psi} k^{p-1}\, d\mathcal{H}^{n-1} &\leq [\mathcal{H}^{n-1}(\Psi)]^{1/p} \left(\int_{\Delta(\hat{w}, s')} k^p d\mathcal{H}^{n-1} \right)^{1-1/p} \\ &\leq c \left[(s')^{1-n}\, \mathcal{H}^{n-1}(\Psi) \right]^{1/p} \int_{\Delta(\hat{w}, s')} k^{p-1} d\mathcal{H}^{n-1} \\ &\leq c^2\, \left[(s')^{1-n} \mathcal{H}^{n-1}(\Psi) \right]^{1/p}\, (s')^{n-p}\, v(a_{s'}(\hat{w}))^{p-1}. \end{aligned} \tag{9.24}$$

Since

$$(s')^{n-p}\, v(a_{s'}(\hat{w}))^{p-1} \approx \int_{\Delta(\hat{w}, s')} k^{p-1}\, d\mathcal{H}^{n-1},$$

we can add (9.23), (9.24) to get after division by $(s')^{n-p}\, v(a_{s'}(\hat{w}))^{p-1}$ that for some c, depending only on the data and $c_\star$,

$$c^{-1} \leq [(s')^{1-n} \mathcal{H}^{n-1}(\Psi)]^{1/p} + (1/\hat{c})^{p-1}. \tag{9.25}$$

Clearly, (9.25), implies (9.22) for $\hat{c}$ large enough with $\hat{K}$ replaced by Ψ. A standard measure theory argument then shows that we can replace Ψ by suitable $\hat{K}$ compact, $\hat{K} \subset \Psi$. Thus (9.22) is valid for $\hat{c}$ large enough.

Next let $K_1 = \hat{K}$ and let D_1 denote the interior of the domain obtained from drawing all line segments from points in $\bar{B}(\hat{w}', s')$ to points in K_1. From (9.21)-(9.22) and (9.5)-(9.6), we conclude for some $\breve{c} \geq 1$, depending only on $c_\star$ and the data that

$$\breve{c} |\nabla v(x)| \geq s^{-1}\, v(\hat{w}') \quad \text{whenever } x \in D_1. \tag{9.26}$$

If

$$\frac{\mathcal{H}^{n-1}(\partial D_1 \cap \Delta(\hat{w}, s'))}{\mathcal{H}^{n-1}(\Delta(\hat{w}, s'))} \leq 7/8, \tag{9.27}$$

choose $c_1 > 2$, depending only on the starlike Lipschitz constant for D, so that if $s'' = (1 - 2/c_1)s'$, then

$$\frac{\mathcal{H}^{n-1}(\Delta(\hat{w}, s''))}{\mathcal{H}^{n-1}(\Delta(\hat{w}, s'))} \geq 99/100. \tag{9.28}$$

Since $\partial D_1 \cap \Delta(\hat{w}, s') = K_1$ it follows from (9.27) (9.28), that there exists $y \in \Delta(\hat{w}, s'')\setminus K_1$. We can apply the argument leading to (9.22), (9.26) with $\hat{w}, s$ replaced by $y, \tau = d(y, K_1)/c_1$. If $\tau' = \tau/c'$, we obtain a compact set

$$\hat{K}(y) \subset \Delta(y, \tau') \subset \Delta(\hat{w}, s')$$

and corresponding starlike Lipschitz domain $\hat{D}(y)$ with center at y' where y' is the point on the ray from z to y with $|y - y'| = \tau/100$. Also $\hat{D}(y)$ is the interior of the set obtained by drawing all rays from $\bar{B}(y', \tau')$ to points in $\hat{K}(y)$ so that $\partial \hat{D}(y) \cap \Delta(y, \tau') = \hat{K}(y)$. Moreover,

$$\begin{aligned} &(+) \quad |\nabla v(x)| \geq c_{\star\star}^{-1}\, d(y, K_1)^{-1}\, v(y') \quad \text{whenever } x \in \hat{D}(y), \\ &(++) \quad \hat{c}\,\mathcal{H}^{n-1}(\partial \tilde{D}(y) \cap \Delta(y, \tau')) \geq (\tau')^{n-1} \end{aligned} \tag{9.29}$$

where $\hat{c}$ is the constant in (9.22) and $c_{\star\star} \geq 1$ depends only on the data and $c_\star$ in (9.6). We now use a Vitalli type covering argument to get $y_1, y_2, \ldots, y_l$, for some positive integer l, satisfying the above with $y = y_i, 1 \leq i \leq l$, and corresponding $\tau_i, \tau_i', y_i', \hat{K}(y_i), \hat{D}(y_i)$. Then (9.29) holds with y replaced by $y_i, 1 \leq i \leq l$, and

$$\frac{\mathcal{H}^{n-1}(\bigcup_{i=1}^{l} \hat{K}(y_i))}{\mathcal{H}^{n-1}(\Delta(\hat{w}, s'))} \geq c^{-1} \tag{9.30}$$

for some $c \geq 1$ depending only on $c_\star$ and the data. Let D_2 be the starlike Lipschitz domain with center at $\hat{w}'$ which is the interior of the domain obtained by drawing all rays from points in $\bar{B}(\hat{w}', \hat{s}')$ to points in

$$K_2 = K_1 \cup (\bigcup_{i=1}^{l} \hat{K}(y_i)).$$

We claim that

$$|\nabla v(x)| \geq \frac{1}{c_-} \frac{v(\hat{w}')}{s} \quad \text{whenever } x \in D_2 \tag{9.31}$$

where c_- depends only on the data. To prove this claim, given $x \in D_2$, let $\hat{x}$ be the point in $\partial D_2 \cap \Delta(\hat{w}, \hat{s}')$ which lies on the line from $\hat{w}'$ through x. If $\hat{x} \in K_1$ it follows from (9.26) that (9.31) is true for suitably large c_-. Otherwise, suppose $\hat{x} \in \hat{K}(y_j)$. If $|\hat{x} - x| \leq \tau_j$ we observe from our construction that there exists $x^* \in \hat{D}(y_j)$ with

$$|\hat{x} - x| \approx |\hat{x} - x^*| \approx |x - x^*| \tag{9.32}$$

where all constants in the ratios depend only on $c_\star$ and the data. Using (9.32), (9.29) with $y = y_j$, (9.6), and Harnack's inequality we deduce that (9.31) holds. If $|\hat{x} - x| > \tau_j$ we can choose x^* in D_1 so that (9.32) is true. Applying (9.26) and arguing as above we get (9.31) once again. This proves our claim in (9.31).

From disjointness of K_1 and $\cup_i \hat{K}(y_i)$ as well as (9.30) it follows that

$$\frac{\mathcal{H}^{n-1}(K_2)}{\mathcal{H}^{n-1}(\Delta(\hat{w}, s'))} \geq c^{-1} + \frac{\mathcal{H}^{n-1}(K_1)}{\mathcal{H}^{n-1}(\Delta(\hat{w}, s'))}. \tag{9.33}$$

Continuing this argument at most N times, where N depends only on the data and $c_\star$ we see from (9.33) that we eventually obtain K_N a compact set $\subset \Delta(\hat{w}', s')$ and D_N a starlike Lipschitz domain with center at $\hat{w}'$ corresponding to K_N for which

$$\frac{\mathcal{H}^{n-1}(K_N)}{\mathcal{H}^{n-1}(\Delta(\hat{w}, s'))} \geq 7/8. \tag{9.34}$$

Also (9.31) is valid for large c_- with D_2 replaced by D_N.

To complete the construction of $\tilde{D}$, we need to estimate $|\nabla v|$ from above. For this purpose let $\mathcal{M}_1 k$ be as in (9.18) with $\tilde{r}$ replaced by s'. Once again we use the Hardy-Littlewood Maximal theorem and also (9.5), (9.6), to find K^* compact $\subset \Delta(\hat{w}, s')$ and $\bar{c} \geq 1$ depending on $c_\star$ and the data such that

$$\mathcal{M}_1 k \leq \bar{c}\,(s')^{1-p}\, v(a_{s'}(\hat{w}))^{p-1} \quad \text{on} \quad K^* \tag{9.35}$$

and

$$\mathcal{H}^{n-1}(K^*) \geq \tfrac{7}{8}\, \mathcal{H}^{n-1}(\Delta(\hat{w}, s')). \tag{9.36}$$

Let D^* be the interior of the domain obtained from drawing all rays from points in $\bar{B}(\hat{w}', s')$ to points in K^*. If $x \in D^*$ then from (9.35), (9.36), (9.6), (9.5), and Harnack's inequality for v, we find for some $\tilde{c}$ that

$$|\nabla v(x)| \leq \tilde{c}\, v(\hat{w}')/s \quad \text{whenever} \quad x \in D^*. \tag{9.37}$$

Let $\tilde{D} = D^* \cap D_N$. From (9.21) it is easily seen for c' large enough that $\tilde{D}$ is starlike Lipschitz with center at w' and starlike Lipschitz constant $\leq c\,(\|\log \hat{\mathcal{R}}\|_{\mathbb{S}^{n-1}} + 1)$. Also, from (9.34), (9.31) with D_2 replaced by D_N, (9.36), and (9.37) we see that (9.19), (9.20) are valid. The proof of Lemma 9.6 is now complete. □

We use Lemmas 9.5 and 9.6 to prove

PROPOSITION 9.7. *Let $D, z, c_\star, b, v, \tau, r, k, w$, be as in Lemma 9.6. Then*

$$\lim_{\substack{x \to y \\ x \in \Gamma(y) \cap B(w, 2r)}} \nabla v(x) \stackrel{def}{=} \nabla v(y) \text{ exists for } \mathcal{H}^{n-1}\text{-a.e } y \in \Delta(w, 2r). \tag{9.38}$$

Moreover, $\Delta(w, 2r)$ has a tangent plane for $\mathcal{H}^{n-1}$ almost every $y \in \Delta(w, 2r)$. If $\mathbf{n}(y)$ denotes the unit normal to this tangent plane pointing into $D \cap B(w, 2r)$, then

$$k(y)^{p-1} = p\frac{f(\nabla v(y))}{|\nabla v(y)|} \tag{9.39}$$

and

$$\nabla v(y) = |\nabla v(y)|\, \mathbf{n}(y) \quad \mathcal{H}^{n-1}\text{-a.e. on } \Delta(w, 2r). \tag{9.40}$$

PROOF. In the proof of Proposition 9.7 we argue as in [**LN3**, Lemma 3.2]. The proof is by contradiction.

Suppose there exists a Borel set $V \subset \Delta(w, 2r)$ with $\mathcal{H}^{n-1}(V) > 0$, such that Proposition 9.7 is false for each $y \in V$. Under this assumption, we let $\hat{w} \in V$ be a point of density for V with respect to $\mathcal{H}^{n-1}|_{\partial D}$. Then

$$\frac{\mathcal{H}^{n-1}(\Delta(\hat{w}, t) \setminus V)}{\mathcal{H}^{n-1}(\Delta(\hat{w}, t))} \to 0 \quad \text{as} \quad t \to 0,$$

and so there exists $c \geq 1$ depending only on $c_\star$ in (9.6) and the data such that

$$c\, \mathcal{H}^{n-1}(\partial \tilde{D} \cap \Delta(\hat{w}, s) \cap V) \geq s^{n-1}$$

provided $s > 0$ is small enough, where $\tilde{D} = \tilde{D}(\hat{w}, s) \subset D$ is the starlike Lipschitz domain defined in Lemma 9.6. To get a contradiction, we show that

$$\text{Proposition 9.7 is true for almost every } y \in \partial \tilde{D} \cap \Delta(\hat{w}, s). \tag{9.41}$$

To do this, let Ψ be the set of all $y \in \partial\tilde{D} \cap \Delta(\hat{w}, s)$ satisfying

$$\begin{aligned} &(a) \quad y \text{ is a point of density for } \Psi \text{ relative to } \mathcal{H}^{n-1}|_{\partial D},\ \mathcal{H}^{n-1}|_{\partial\tilde{D}}, \tau, \\ &(b) \quad \text{There is a tangent plane } T(y) \text{ to both } \partial D, \partial\tilde{D} \text{ at } y, \\ &(c) \quad \lim_{t\to 0} t^{1-n}\mathcal{H}^{n-1}(\partial D \cap B(y,t)) = \lim_{t\to 0} t^{1-n}\mathcal{H}^{n-1}(\partial\tilde{D} \cap B(y,t)) = b', \\ &(d) \quad \lim_{t\to 0} t^{1-n}\tau(\partial D \cap B(y,t)) = b'\, k(y)^{p-1}. \end{aligned} \tag{9.42}$$

In (9.42), b' denotes the Lebesgue $(n-1)$-measure of the unit ball in $\mathbb{R}^{n-1}$.

We claim that

$$\mathcal{H}^{n-1}(\partial\tilde{D} \cap \Delta(\hat{w}, s) \setminus \Psi) = 0. \tag{9.43}$$

Indeed (a) of (9.42) for $\mathcal{H}^{n-1}$-almost every y is a consequence of the fact that $\mathcal{H}^{n-1}|_{\partial D}$, $\mathcal{H}^{n-1}|_{\partial\tilde{D}}$ are regular Borel measures and differentiation theory while (a) of this display for τ and $\mathcal{H}^{n-1}|_{\partial D}$ for almost every y, follows from the same observations and Lemma 9.5. (9.42) (b) follows from the Lipschitz character of $D, \tilde{D}$, and Rademacher's theorem ([**EG**, Chapter 3]). Finally (c) and (d) of this display are consequences of the Lebesgue differentiation theorem and Lemma 9.5. Thus, (9.43) is true.

We now use a blowup argument to complete the proof of Proposition 9.7. Let Ψ, s be as above and $y \in \Psi$. Since $\mathcal{A}$-harmonic functions are invariant under translation we may assume that $y = 0$. Let $\{t_m\}_{m\geq 1}$ be a decreasing sequence of positive numbers with limit zero and $t_1 << s$. Let

$$\begin{aligned} D_m &= \{x : \ t_m x \in D \cap B(\hat{w}, s)\}, \\ \tilde{D}_m &= \{x : \ t_m x \in \tilde{D} \cap B(\hat{w}, s)\}, \\ v_m(x) &= t_m^{-1}\, v(t_m x) \quad \text{whenever} \quad t_m x \in B(\hat{w}, s). \end{aligned}$$

Fix $R >> 1$. Then for m sufficiently large, say $m \geq m_0, m_0 = m_0(R)$, we note that v_m is $\mathcal{A}$-harmonic in $D_m \cap B(0, 2R)$ and continuous in $B(0, 2R)$ with $v_m \equiv 0$ on $B(0, 2R) \setminus D_m$. Let

$$\nu_m(J) = t_m^{1-n}\tau(t_m J) \quad \text{whenever } J \text{ is a Borel subset of } B(0, 2R). \tag{9.44}$$

Then ν_m is the measure corresponding to v_m on $B(0, 2R)$, as in Lemma 9.4 for $m \geq m_0$. Let $\xi \in \mathbb{S}^{n-1}$ be a normal to $T(0)$. We assume as we may that $H = \{x : \langle x, \xi\rangle > 0\}$ contains $\hat{w}'$. Then from Lipschitz starlikeness of $D, \tilde{D}$, and (9.42) (b) we deduce that

$$\begin{aligned} d_{\mathcal{H}}(D_m &\cap B(0,R), H \cap B(0,R)) \\ &+ d_{\mathcal{H}}(\tilde{D}_m \cap B(0,R), H \cap B(0,R)) \to 0 \quad \text{as} \quad m \to \infty, \end{aligned} \tag{9.45}$$

where $d_{\mathcal{H}}$ as defined in chapter 2 denotes Hausdorff distance. Let $\eta = v(\hat{w}')/s$ and from (9.20) we see that

$$|\nabla v_m| \leq c\eta \quad \text{on } \tilde{D}_m. \tag{9.46}$$

Also, from (9.45), (9.46), (9.3) for v_m, and (9.6) we deduce that

$$|v_m(x)| \leq c\left(\frac{d(x, \partial D_m)}{R}\right)^{\beta} \eta\, R \quad \text{whenever } x \in D_m \cap B(0, R), \tag{9.47}$$

where β is the Hölder exponent in (9.3). From (9.46), (9.47), and (3.2), we see that a subsequence of $\{v_m\}$, say $\{v'_m(x)\}$ where $v'_m(x) = v(t'_m x)/t'_m$, converges uniformly on compact subsets of $\mathbb{R}^n$ to a Hölder continuous function $\hat{v}$ with $\hat{v} \equiv 0$ in $\mathbb{R}^n \setminus H$. Also $\hat{v} \geq 0$ is $\mathcal{A} = \nabla f$-harmonic in H.

We now apply a boundary Harnack inequality in Theorem 1 of [**LLN**] with Ω, u replaced by $H, \langle x, \xi \rangle^+$, respectively. Letting $r \to \infty$ in this inequality, we get $\hat{v}(x) = \gamma \langle x, \xi \rangle^+$ for some $\gamma \geq 0$, where $C^+ = \max(C, 0)$. Let ν'_m be the measure corresponding to v'_m and observe from (9.5), (9.47) that the sequence of measures, $\{\nu'_m\}_{m\geq 1}$, corresponding to $\{v'_m\}_{m \geq 1}$, have uniformly bounded total masses on $B(0, R)$. Also from (9.2)-(9.4), (9.47), we see that $\{v'_m\}$ is uniformly bounded in $W^{1,p}(B(0, R))$. Using these facts and (3.2) we obtain that

$$\{\nu'_m\} \text{ converges weakly to } \nu \quad \text{as } m \to \infty$$

where ν is the measure associated with $\gamma \, \langle x, \xi \rangle^+$. Using integration by parts and the fact that $\langle x, \xi \rangle^+$ is $\mathcal{A} = \nabla f$-harmonic in H we get

$$d\nu = \gamma^{p-1} \langle \nabla f(\xi), \xi \rangle d\mathcal{H}^{n-1} \left.\right|_{\partial H} = p\gamma^{p-1} f(\xi) d\mathcal{H}^{n-1} \left.\right|_{\partial H}$$

where we have also used p-homogeneity of f. From this computation, weak convergence, (9.44), and (9.42) (d), we have

$$\begin{aligned} p\gamma^{p-1} \, f(\xi) \, b' \, R^{n-1} &= \lim_{m\to\infty} \nu'_m(B(0, R)) \\ &= \lim_{m\to\infty} (t'_m)^{1-n} \nu(B(0, Rt'_m)) \\ &= b' \, R^{n-1} k^{p-1}(0). \end{aligned} \tag{9.48}$$

Also, from our earlier observations we see that $x \mapsto t^{-1} v(tx)$ converges uniformly as $t \to 0$ to $\gamma \langle x, \xi \rangle^+$ on compact subsets of $\mathbb{R}^n$ and $x \mapsto \nabla v(tx)$ converges uniformly to $\gamma \xi$ as $t \to 0$ when x lies in a compact subset of H. Given $0 < \theta < 1$, let

$$K_\theta = \{x \in H : \ \langle x, \xi \rangle \geq \theta |x|\}.$$

In view of these remarks we conclude that

$$\lim_{t\to 0} \nabla v(t\omega) = \gamma \xi \tag{9.49}$$

whenever $0 < \theta < 1$ is fixed and $\omega \in K_\theta$ with $|\omega| = 1$. It is easily seen for given $0 < b < 1$ and $t > 0$ small that there exists $\theta > 0$ such that $\Gamma(0)$ defined relative to D and b satisfies $\Gamma(0) \cap B(0, t) \subset K_\theta$. From this observation and (9.49) we conclude the validity of (9.38) independently of b. Then $\gamma \xi = \nabla v(0)$ by definition so using (9.48) to solve for $k(0)$ we arrive at (9.39) and (9.40). This completes the proof of (9.41) which as mentioned above this display gives a contradiction to our assumption that Proposition 9.7 is false. □

CHAPTER 10

Boundary Harnack inequalities

In this chapter we use our work in chapter 9 to prove boundary Harnack inequalities for the ratio of two $\mathcal{A} = \nabla f$-harmonic functions $\tilde{u}, \tilde{v}$, which are $\mathcal{A}$-harmonic in $B(w, 4r) \cap D'$ and continuous in $B(w, 4r)$ with $\tilde{u} = \tilde{v} \equiv 0$ on $B(w, 4r) \setminus D'$. Here D' is a bounded Lipschitz domain.

To set the stage for these inequalities, let D be a starlike Lipschitz domain with center z. Let

$$w \in \partial D \text{ and } \ 0 < r < |w - z|/100.$$

Let $\breve{v}_i > 0$, for $i = 1, 2$, be $\mathcal{A} = \nabla f$-harmonic functions satisfying (9.6) in $B(w, 4r) \cap D$. Assume also that $\breve{v}_i$ is continuous in $B(w, 4r)$ with $\breve{v}_i \equiv 0$ on $B(w, 4r) \setminus D$, for $i = 1, 2$.

Let $\hat{w} \in \partial D \cap B(w, r), 0 < s < r$, and let $\tilde{D}_i = \tilde{D}_i(\hat{w}, s)$ be the starlike Lipschitz domains in Lemma 9.6 with center at $\hat{w}'$ defined relative to $\breve{v}_i$ and D for $i = 1, 2$. Put $\tilde{D} = \tilde{D}_1 \cap \tilde{D}_2$.

From this lemma we see that

$$|\nabla \breve{v}_i(x)| \approx \breve{v}_i(\hat{w}')/s \text{ when } x \in \tilde{D} \text{ and } i = 1, 2, \tag{10.1}$$

where ratio constants depend only on the data and $c_\star$. Also if $s' = s/c'$, then

$$\frac{\mathcal{H}^{n-1}[\partial \tilde{D} \cap \partial D \cap B(\hat{w}, s')]}{\mathcal{H}^{n-1}[\partial D \cap B(\hat{w}, s')]} \geq 1/2. \tag{10.2}$$

Given $t_1, t_2 \geq 0$, and for $y \in \tilde{D}$ set

$$d\tilde{\gamma}(y) := \left[d(y, \partial \tilde{D}) \max_{x \in B(y, \frac{1}{4} d(y, \partial \tilde{D}))} \mathcal{M}(x) \right] dy$$

where

$$\mathcal{M}(x) = \left\{ [t_1 |\nabla \breve{v}_1(x)| + t_2 |\nabla \breve{v}_2(x)|]^{2p-6} \sum_{i,j=1}^{n} \left[t_1 |(\breve{v}_1(x))_{x_i x_j}| + t_2 |(\breve{v}_2(x))_{x_i x_j}| \right]^2 \right\}.$$

We show $\tilde{\gamma}$ is a Carleson measure on $\tilde{D}$. More specifically, we prove the following lemma.

Lemma 10.1. *With the above notation, if $\hat{x} \in \partial \tilde{D}$ and $0 < \rho \leq \operatorname{diam}(\tilde{D})$, then*

$$\tilde{\gamma}(\tilde{D} \cap B(\hat{x}, \rho)) \leq c \left(t_1 \frac{\breve{v}_1(\hat{w}')}{s} + t_2 \frac{\breve{v}_2(\hat{w}')}{s} \right)^{2p-4} \rho^{n-1}$$

where c depends only on $c_\star$ and the data.

PROOF. Observe from (10.1) and (3.3) that if

$$\mathcal{I} = \left[\frac{(t_1\breve{v}_1(\hat{w}') + t_2\breve{v}_2(\hat{w}'))}{s'}\right]^{2p-6}$$

then

(10.3)

$$\begin{aligned}
\tilde{\gamma}(\tilde{D}\cap B(\hat{x},\rho)) &= \int_{\tilde{D}\cap B(\hat{x},\rho)} d\tilde{\gamma}(y)\\
&\le c\mathcal{I}\int_{\tilde{D}\cap B(\hat{x},\rho)} d(y,\partial\tilde{D}) \max_{B(y,\frac{1}{4}d(y,\partial\tilde{D}))}\left\{\sum_{i,j=1}^{n}(t_1|(\breve{v}_1)_{x_ix_j}| + t_2|(\breve{v}_2)_{x_ix_j}|)^2\right\} dy\\
&\le c^2\mathcal{I}\int_{\tilde{D}\cap B(\hat{x},\rho)} d(y,\partial\tilde{D})^{1-n}\left(\int_{B(y,\frac{3}{4}d(y,\partial\tilde{D}))}\sum_{i,j=1}^{n}(t_1|(\breve{v}_1)_{x_ix_j}| + t_2|(\breve{v}_2)_{x_ix_j}|)^2 dx\right) dy\\
&\le c^3\mathcal{I}\int_{\tilde{D}\cap B(\hat{x},\rho)} d(y,\partial\tilde{D})\sum_{i,j=1}^{n}(t_1|(\breve{v}_1)_{x_ix_j}(y)| + t_2|(\breve{v}_2)_{x_ix_j}(y)|)^2\, dy = \mathcal{II},
\end{aligned}$$

where to get the last integral we have interchanged the order of integration in the second integral. From (9.6) and (10.1) we find for $y\in\tilde{D}$, and $i=1,2$, that

$$d(y,\partial\tilde{D}) \le d(y,\partial D) \le c(s/\breve{v}_i(\hat{w}'))\,\breve{v}_i(y). \tag{10.4}$$

Using (10.4) in (10.3), it follows that if

$$\mathcal{II}_i = t_i^2\,(s/\breve{v}_i(\hat{w}'))\,\mathcal{I}\int_{\tilde{D}\cap B(\hat{x},\rho)} \breve{v}_i(y)\sum_{j,k=1}^{n}|(\breve{v}_i)_{x_jx_k}(y)|^2 dy \quad \text{for } i=1,2$$

then

$$\mathcal{II} \le \tilde{c}\,(\mathcal{II}_1 + \mathcal{II}_2) \tag{10.5}$$

where $\tilde{c}$ depends only on $c_\star$ in (9.6) and the data.

To estimate $\mathcal{II}_i$ for $i=1,2$, fix $i\in\{1,2\}$ and for small $\delta>0$, put

$$\vartheta = \delta^{-1}\nabla\breve{v}_i(x+\delta e_l) \quad \text{and} \quad \upsilon = \delta^{-1}\nabla\breve{v}_i(x).$$

By repeating the argument from (5.7) to (5.9) and letting $\delta\to 0$ in this equality we deduce from (10.1), (3.3) that if $\zeta = (\breve{v}_i)_{x_l}, 1\le l\le n$, then ζ is a weak solution in $\tilde{D}$ to

$$\mathcal{L}_i\zeta = \sum_{k,j=1}^{n}\left((\breve{b}_i)_{kj}\zeta_{x_j}\right)_{x_k} = 0 \tag{10.6}$$

where

$$(\breve{b}_i)_{kj}(x) = f_{\eta_k\eta_j}(\nabla\breve{v}_i(x)) \quad \text{for } 1\le k,j\le n. \tag{10.7}$$

Also $\breve{v}_i$ is a solution to (10.6) as follows from $\mathcal{A}$-harmonicity of $\breve{v}_i$ and p-homogeneity of f. Using (10.6), (10.7), the structure assumptions on $\mathcal{A}$, and (10.1) we deduce for $i = 1, 2$, that

$$
\begin{aligned}
\mathcal{L}_i(|\nabla \breve{v}_i(x)|^2) &\geq 2 \sum_{k,j,l=1}^{n} (\breve{b}_i)_{kj}(x)[(\breve{v}_i)_{x_k x_l}(\breve{v}_i)_{x_j x_l}] \\
&\geq c^{-1}|\nabla \breve{v}_i(\hat{w}')|^{p-2} \sum_{k,j=1}^{n} [(\breve{v}_i)_{x_k x_j}]^2 \\
&\geq c^{-2}(\breve{v}_i(\hat{w}')/s)^{p-2} \sum_{k,j=1}^{n} [(\breve{v}_i)_{x_k x_j}]^2
\end{aligned} \tag{10.8}
$$

weakly in $\tilde{D}$. Given $t \in (1/2, 1)$ and $y \in \partial\tilde{D}$, let $y(t)$ be that point on the line segment from $\hat{w}'$ to y with $|y(t) - \hat{w}'| = t|y - \hat{w}'|$. Let $\tilde{D}(t)$ be the union of all half open line segments $[\hat{w}', y(t))$ joining $\hat{w}'$ to $y(t)$ when $y \in \partial\tilde{D}$. Using starlike Lipschitzness of $\partial\tilde{D}(t)$, the fact that

$$
\breve{v}_i\mathcal{L}_i(|\nabla \breve{v}_i|^2) = \breve{v}_i\mathcal{L}_i(|\nabla \breve{v}_i|^2) - \mathcal{L}_i(\breve{v}_i)|\nabla \breve{v}_i|^2
$$

weakly in $\tilde{D}(t)$, (10.1), (10.8), and integration by parts we obtain for $\mathcal{H}^1$ almost every $t \in (1/2, 1)$ that

(10.9)

$$
\begin{aligned}
&(\breve{v}_i(\hat{w}')/s)^{p-2} \int_{\tilde{D}(t)\cap B(\hat{x},\rho)} \sum_{k,j=1}^{n} \breve{v}_i\, |(\breve{v}_i)_{x_k x_j}|^2 \\
&\leq c \left| \int_{\partial\,[\tilde{D}(t)\cap B(\hat{x},\rho)]} \sum_{k,j=1}^{n} (\breve{b}_i)_{kj}[(\breve{v}_i\,(|\nabla \breve{v}_i|^2)_{x_k} - |\nabla \breve{v}_i|^2(\breve{v}_i)_{x_k})]\nu(t)_j d\mathcal{H}^{n-1} \right|
\end{aligned}
$$

where $\nu(t)$ denotes the unit outer normal to $\tilde{D}(t) \cap B(\hat{x}, \rho)$ and $c \geq 1$ depends only on $c_\star$ and the data. Using once again (10.1) and (3.3) we can estimate the right-hand side of (10.9). Doing this and using the resulting estimate in (10.9) we deduce that

$$
\int_{\tilde{D}(t)\cap B(\hat{x},\rho)} \sum_{k,j=1}^{n} \breve{v}_i\, |(\breve{v}_i)_{x_k x_j}|^2\, dx \,\leq \breve{c} \left(\frac{\breve{v}_i(\hat{w}')}{s}\right)^3 \rho^{n-1} \tag{10.10}
$$

where $\breve{c}$ depends only on $c_\star$ and the data. Letting $t \to 1$ and using Fatou's lemma we see that (10.10) remains valid with $\tilde{D}(t)$ replaced by $\tilde{D}$. In view of (10.10) for $t = 1$ and (10.5) we conclude first that

$$
\begin{aligned}
\mathcal{II} &\leq \tilde{c}\mathcal{I} \left[t_1^2 \left(\frac{\breve{v}_1(\hat{w}')}{s}\right)^2 + t_2^2 \left(\frac{\breve{v}_2(\hat{w}')}{s}\right)^2 \right] \rho^{n-1} \\
&\leq \bar{c} \left[t_1 \frac{\breve{v}_1(\hat{w}')}{s} + t_2 \frac{\breve{v}_2(w')}{s} \right]^{2p-4} \rho^{n-1}
\end{aligned}
$$

and thereupon from (10.3) and arbitrariness of $\hat{x}, \rho$ that Lemma 10.1 is true. □

We continue under the assumption that $\breve{v}_i, D, t_i, i = 1, 2, s, \hat{w}, \hat{w}', \tilde{D}$ are as in Lemma 10.1. Let $\breve{v} = t_1\breve{v}_1 - t_2\breve{v}_2$. Using (5.7) with $\vartheta = t_1\nabla\breve{v}_1$, $\upsilon = t_2\nabla\breve{v}_2$, $\mathcal{A}$-harmonicity of $\breve{v}_i$, and p-homogeneity of f, we deduce as in (10.6) that $\breve{v}$ is a weak solution in D to

$$\breve{\mathcal{L}}\breve{v} = \sum_{k,j=1}^{n} \left(\tilde{b}_{kj}\breve{v}_{x_j}\right)_{x_k} = 0 \tag{10.11}$$

where at x

$$\tilde{b}_{kj}(x) = \int_0^1 f_{\eta_k\eta_j}(st_1\nabla\breve{v}_1(x) + (1-s)t_2\nabla\breve{v}_2(x))ds \quad \text{for } 1 \leq k, j \leq n. \tag{10.12}$$

Now, if

$$\beta(x) = (t_1|\nabla\breve{v}_1(x)| + t_2|\nabla\breve{v}_2(x)|)^{p-2}$$

then

$$\sum_{k,j=1}^{n} \tilde{b}_{kj}(x)\xi_k\xi_j \approx \beta(x)|\xi|^2 \quad \text{whenever} \quad \xi \in \mathbb{R}^n \setminus \{0\}. \tag{10.13}$$

Ratio constants depend only on the data. We note from (9.20) for $\breve{v}_1, \breve{v}_2$ that

$$\beta(x) \approx (t_1\breve{v}_1(\hat{w}')/s + t_2\breve{v}_2(\hat{w}')/s)^{p-2} = \phi \tag{10.14}$$

where $\phi > 0$ when $x \in \tilde{D}$. Thus $(\phi^{-1}\tilde{b}_{kj})$ is uniformly elliptic in $\tilde{D}$ with ellipticity constant ≈ 1. It is then classical (see [**LSW**, Theorem (6.1)]) that Green's function for this operator with pole at $\hat{w}' \in \tilde{D}$ exists as well as the corresponding elliptic measure, $\tilde{\omega}(\cdot, \hat{w}')$. Moreover, as in [**CFMS**, section 4] there exists $\bar{c} \geq 1$ depending only on $c_\star$ and the data such that

$$\bar{c}\,\tilde{\omega}(\partial\tilde{D} \cap B(\hat{w}, s'), \hat{w}') \geq 1. \tag{10.15}$$

Using Lemma 10.1 we see that Theorem 2.6 in [**KP**] can be applied to conclude that $\tilde{\omega}(\cdot, \hat{w}')$ is an A^∞-weight on $\partial\tilde{D}$ in the following sense. There exists $\tilde{c}_+ \geq 1$ depending only on $c_\star$ and the data such that if $\hat{x} \in \partial\tilde{D}, 0 < \rho < \operatorname{diam} \tilde{D}$, and $K \subset \partial\tilde{D} \cap B(\hat{x}, \rho)$ is a Borel set then

$$\frac{\mathcal{H}^{n-1}(K)}{\mathcal{H}^{n-1}(\partial\tilde{D} \cap B(\hat{x}, \rho))} \geq 1/4 \quad \Rightarrow \quad \frac{\tilde{\omega}(K, \hat{w}')}{\tilde{\omega}(\partial\tilde{D} \cap B(\hat{x}, \rho), \hat{w}')} \geq c_+^{-1}. \tag{10.16}$$

To use this result and avoid existence questions for elliptic measure as well as the Green's function defined relative to $(\phi^{-1}\tilde{b}_{kj})$ in D we temporarily assume that

$$\mathcal{R} \in C^\infty(\mathbb{S}^{n-1}) \quad \text{where} \quad \partial D = \{z + \mathcal{R}(\zeta)\,\zeta : \ \zeta \in \mathbb{S}^{n-1}\} \tag{10.17}$$

and $\mathcal{R}$ is as in Definition 9.1. Then from [**Li**, Theorem 1] it follows that $\nabla\breve{v}_i, i = 1, 2$, has a nonzero locally Hölder continuous extension to $\bar{D} \cap B(w, 4r)$. From this theorem and (10.13) we deduce that if $y \in D \cap B(w, 2r)$ and $0 < s < r$, then Green's function for $\breve{\mathcal{L}}$ with pole at y and the corresponding elliptic measure $\omega(\cdot, y)$, exist relative to $D \cap B(w, 2r)$. Then from (10.2), (10.15), (10.16) with $\hat{x} = \hat{w}, \rho = s', K = \partial\tilde{D} \cap \partial D \cap B(\hat{w}, s')$ and the weak maximum principle for $\breve{\mathcal{L}}$ we find $c_{++} \geq 1$, depending only on $c_\star$ and the data such that

$$\begin{aligned} \omega(\partial D \cap B(\hat{w}, s'), \hat{w}') &\geq \tilde{\omega}(\partial D \cap \partial\tilde{D} \cap B(\hat{w}, s'), \hat{w}') \\ &\geq c_{++}^{-1}. \end{aligned} \tag{10.18}$$

From arbitrariness of $\hat{w}, s$, we deduce from (10.18) and a covering argument that if $\hat{x} \in \partial D \cap B(w, r)$, $0 < t \leq r/2$, then

$$\omega(\partial D \cap [B(\hat{x}, 7t/8) \setminus B(\hat{x}, 5t/8)], \cdot) \geq c^{-1} \quad \text{on} \quad \partial B(\hat{x}, 3t/4) \cap D \tag{10.19}$$

where c depends only on $c_\star$ and the data. Let $g(\cdot,\zeta)$ denote Green's function for $\breve{\mathcal{L}}$ in $D \cap B(\hat{x},t)$ with pole at ζ in $D \cap B(\hat{x},t)$. Let $\hat{x}'$ denote the point on the ray from z to $\hat{x}$ with $|\hat{x}-\hat{x}'| = t/100$. Let c' be as in Lemma 9.6. We assume as we may that if $t' = t/c'$, then $B(\hat{x}', 2t') \subset D \cap B(\hat{x},t)$. Let $a = \max_{\partial B(\hat{x}',t')} g(\cdot,\hat{x}')$. Let $t'' = t/\hat{c}^+, c' << \hat{c}^+$, and suppose $y \in D \cap B(\hat{x},t'')$. Choose $\bar{y} \in \partial D \cap B(\hat{x}, 2t'')$ with $|\bar{y}-y| = d(y,\partial D)$. Then using the iteration argument in **[LN]** from (3.16) to (3.26) it follows that for $\hat{c}^+ \geq 100c'$, large enough, depending only on the data and $c_\star$, that

$$\omega(\partial D \cap [B(\bar{y},t) \setminus B(\bar{y},t/2)], y) \leq \hat{c}^+ \frac{g(y,\hat{x}')}{a}$$

and

$$\partial D \cap [B(\hat{x},7t/8) \setminus B(\hat{x},5t/8)] \subset \partial D \cap [B(\bar{y},t) \setminus B(\bar{y},t/2)]$$

whenever $y \in \partial D \cap B(\hat{x},t'')$.

Thus from the maximum principle for solutions to $\breve{\mathcal{L}}$,

$$\omega(\partial D \cap [B(\hat{x},7t/8) \setminus B(\hat{x},5t/8)], \cdot) \leq \hat{c}^+ \frac{g(\cdot,\hat{x}')}{a} \quad \text{on } D \cap B(\hat{x},t''). \tag{10.20}$$

Using (10.20) we prove

LEMMA 10.2. *Let $\breve{v}_i$, for $i = 1, 2, D, r, w$ be as introduced above Lemma 10.1 and suppose (10.17) holds. Let $\hat{x} \in \partial D \cap B(w,r), 0 < t \leq r/2$, and define $\hat{x}'$ relative to $\hat{x}$ as in (10.20). Let $\breve{\mathcal{L}}$ be as in (10.11) and let h_1, h_2 be positive weak solutions to $\breve{\mathcal{L}} h_i = 0$ for $i = 1, 2$, in $D \cap B(\hat{x},t)$. Suppose also that h_1, h_2 are continuous in $\bar{D} \cap \bar{B}(\hat{x},t)$ with boundary value 0 on $\partial D \cap B(\hat{x},t)$. Then for some $\tilde{c} \geq 1$ depending only on $c_\star$ and the data,*

$$\tilde{c}^{-1} \frac{h_1(\hat{x}')}{\max_{\bar{D}\cap \partial B(\hat{x},t)} h_2} \leq \frac{h_1(x)}{h_2(x)} \leq \tilde{c} \frac{\max_{\bar{D}\cap\partial B(\hat{x},t)} h_1}{h_2(\hat{x}')} \tag{10.21}$$

whenever $x \in D \cap B(\hat{x},t'')$. If $t_1 = 1, \breve{v}_2 \equiv 0$, then there exists $c_1' \geq 1, \theta \in (0,1)$, depending only on the data and c_* such that if $t''' = t''/c_1'$, then

$$\left| \frac{h_1(x)}{h_2(x)} - \frac{h_1(y)}{h_2(y)} \right| \leq c \left(\frac{|x-y|}{t} \right)^\theta \frac{h_1(x)}{h_2(x)} \quad \textit{whenever } x, y \in B(\hat{x},t'''). \tag{10.22}$$

PROOF. We note from (10.19) and the maximum principle for solutions to $\breve{\mathcal{L}}$, that

$$h_i \leq c \left(\max_{\bar{D}\cap\partial B(\hat{x},t)} h_i \right) \omega(\partial D \cap [B(\hat{x},7t/8) \setminus B(\hat{x},5t/8)], \cdot) \quad \text{in } D \cap B(\hat{x},3t/4)$$

and

$$h_i \geq c^{-1} h_i(\hat{x}') g(\cdot,\hat{x}')/a \quad \text{in } D \cap B(\hat{x},t) \setminus B(\hat{x}',t')$$

when $i = 1, 2$ for some $c \geq 1$, depending only on the data and $c_\star$. These inequalities and (10.20) give (10.21). To prove (10.22) we note that $\breve{v}_1$ is now a solution to $\breve{\mathcal{L}}$ in $D \cap B(w, 4r)$ so from (9.3) and (10.21) with $h_2 = \breve{v}$ we have

$$h_1(x) \leq c \left(\max_{\bar{D}\cap\partial B(\hat{x},t)} h_1 \right) (|x-\hat{x}|/t)^{\tilde{\beta}} \quad \text{for } x \in B(\hat{x},t''). \tag{10.23}$$

Now (10.23), arbitrariness of $\hat{x}, t$ and the local Harnack inequality implied by (10.13) for h_1 yield as in [**CFMS**] that for some $\beta > 0, c_1' \geq 1$ depending only on the data and $c_\star$, that if $t''' = t''/c_1'$, then

$$h_1(x) \leq c_1'(\rho/t)^\beta h(a_\rho(y)) \quad \text{for } x \in D \cap B(y,\rho), \tag{10.24}$$

whenever $0 < \rho \leq t'''$, $x \in D \cap B(y,\rho)$, and $y \in \partial D \cap B(w,r)$. (10.21) and (10.24) imply that

$$\frac{h_1(x)}{h_2(x)} \approx \frac{h_1(a_{t'''}(\hat{x}))}{h_2(a_{t'''}(\hat{x}))} \quad \text{for } x \in D \cap B(\hat{x}, t''') \tag{10.25}$$

where once again all constants depend only on the data and $c_\star$.

Next if $\zeta \in \partial D \cap B(\hat{x},\rho)$, $0 < \rho \leq t$, we let

$$M(\rho) = \sup_{B(\zeta,\rho)} \frac{h_1}{h_2} \quad \text{and} \quad m(\rho) = \inf_{B(\zeta,\rho)} \frac{h_1}{h_2}$$

Also put

$$\operatorname{osc}(\rho) := M(\rho) - m(\rho) \quad \text{for} \quad 0 < \rho < t.$$

Then, if ρ is fixed we see that $h_1 - h_2 m(\rho)$ and $h_2 M(\rho) - h_1$ are positive solutions to $\breve{\mathcal{L}}$ in $D \cap B(\hat{x},\rho)$ so we can use (10.25) with h_1 replaced by $h_1 - m(\rho)h_2$ to get that if $\rho'''/\rho = t'''/t$, then

$$M(\rho''') - m(\rho) \leq c^*(m(\rho''') - m(\rho)).$$

Likewise using (10.25) with $M(\rho)h_2 - h_1$ replacing h_1, we obtain that

$$M(\rho) - m(\rho''') \leq c^*(M(\rho) - M(\rho''')).$$

Adding these inequalities we obtain after some arithmetic that

$$\operatorname{osc}(\rho''') \leq \frac{c^* - 1}{c^* + 1} \operatorname{osc}(\rho) \tag{10.26}$$

where c^* depends only on $c_\star$ and the data. Iterating (10.26) we conclude for some $c \geq 1, \alpha' \in (0,1)$, depending only on the data and $c_\star$ that

$$\operatorname{osc}(s) \leq c(s/t)^{\alpha'} \operatorname{osc}(t) \quad \text{whenever} \quad 0 < s \leq t \leq r. \tag{10.27}$$

Lemma 10.2 follows from this inequality, arbitrariness of ζ, and the interior Hölder continuity-Harnack inequalities for solutions to $\breve{\mathcal{L}}$. □

We use Lemma 10.2 to prove

LEMMA 10.3. *Let $\breve{v}_i$, for $i = 1, 2$, w, r, D, be as introduced above Lemma* 10.1. *Suppose* (10.17) *holds. There exists $\bar{c}_+ \geq 1$ depending only on $c_\star$ and the data such that if $r^+ = r/\bar{c}_+$ then*

$$\bar{c}_+^{-1} \frac{\breve{v}_1(a_{r^+}(w))}{\breve{v}_2(a_{r^+}(w))} \leq \frac{\breve{v}_1(y)}{\breve{v}_2(y)} \leq \bar{c}_+ \frac{\breve{v}_1(a_{r^+}(w))}{\breve{v}_2(a_{r^+}(w))}, \tag{10.28}$$

whenever $y \in D \cap B(w, r^+)$.

PROOF. Our proof is similar to the proof of Lemma 4.9 in [**LN4**]. To prove the left-hand inequality in (10.28) we set

$$t_1 = \frac{T}{\breve{v}_1(a_{\bar{r}}(w))} \quad \text{and} \quad t_2 = \frac{1}{\breve{v}_2(a_{\bar{r}}(w))}$$

where $\bar{r}$ is as in (9.3). Let $r^+ = \bar{r}''$ where $\bar{r}''$ is as in Lemma 10.2 with $\bar{r} = t$. We also let

$$\breve{v} = t_1\breve{v}_1 - t_2\breve{v}_2 \quad \text{in} \quad D \cap B(w, 4r)$$

where T is to be determined so that $\breve{v} \geq 0$ in $D \cap B(w, r^+)$. Let $\breve{\mathcal{L}}$ be as in (10.11) relative to $\breve{v}$ and let h_1, h_2 be weak solutions to $\breve{\mathcal{L}}$ in $D \cap B(w, \bar{r})$ with continuous boundary values

$$h_i(x) = \frac{\breve{v}_i(x)}{\breve{v}_i(a_{\bar{r}}(w))} \quad \text{whenever } x \in \partial[D \cap B(w, \bar{r})] \text{ for } i = 1, 2.$$

From (9.3) we see for $i = 1, 2$, that

$$\breve{v}_i(a_{\bar{r}}(w)) \approx \max_{D \cap B(w, \bar{r})} \breve{v}_i.$$

Using this inequality and (10.21) we see that if $\bar{w}'$ denotes the point on the line segment from z to w with $|w - \bar{w}'| = \bar{r}/100$, then

$$\frac{h_1}{h_2} \geq \tilde{c}^{-1} h_1(\bar{w}') = T^{-1} \quad \text{on } D \cap B(w, r^+). \tag{10.29}$$

Thus, if T is as in (10.29) then $Th_1 - h_2 \geq 0$ in $D \cap B(w, r^+)$. Also $Th_1 - h_2$ and $\breve{v}$ are weak solutions to $\breve{\mathcal{L}}$ in $D \cap B(w, \bar{r})$ and these functions have the same boundary values, so from the maximum principle for this PDE we have

$$\breve{v} = Th_1 - h_2 \quad \text{in } D \cap B(w, \bar{r}).$$

Thus to complete the proof of the left-hand inequality in (10.28) it suffices to show that

$$h_1(a_{r^+}(w)) \approx h_1(\bar{w}')) \approx 1 \quad \text{and} \quad h_2(a_{r^+}(w)) \approx 1 \tag{10.30}$$

where ratio constants depend only on $c_\star$ and the data. To do this let $\tilde{w}$ be the point on the line segment from z to w which also lies on $\partial B(w, \bar{r})$. Then

$$\breve{v}_i \approx \breve{v}_i(\tilde{w}) \quad \text{in} \quad B(\tilde{w}, d(\tilde{w}, \partial D)/8) \quad \text{and} \quad \breve{v}_i(\tilde{w}) \approx \breve{v}_i(a_{\bar{r}}(w)).$$

From (10.13), the structure assumptions on $\mathcal{A}$ in Definition 2.1, and Lemma 3.2 we see that $\beta(\tilde{w})^{-1}\breve{\mathcal{L}}$ is uniformly elliptic in $B(\tilde{w}, d(\tilde{w}, \partial D)/8)$ with ellipticity constant ≈ 1. Using these facts we can apply estimates for elliptic measure from [**CFMS**] to conclude first that $h_i(\tilde{w}) \approx h_i(w^*), i = 1, 2$, where w^* lies on the line segment from $\tilde{w}$ to w with $d(w^*, \partial[D \cap B(w, \bar{r})]) \approx \bar{r}$. We can then use Harnack's inequality in a chain of disks connecting w^* to $a_{r^+}(w), \bar{w}'$, to eventually conclude (10.30). This proves the left-hand inequality in (10.28). To get the right-hand inequality in (10.28) we argue as above with $\breve{v}_1, \breve{v}_2$ interchanged. Thus, (10.28) is valid. □

Our goal now is to show that Lemmas 10.2, 10.3, remain valid without assumption (10.17) and (9.6) for certain $\breve{v}_1, \breve{v}_2$. To do this we first prove a lemma on the "Green's function" for $\mathcal{A}$-harmonic functions in a bounded domain O with pole at $w \in O$. In this lemma G denotes the fundamental solution for $\mathcal{A}$-harmonic functions with pole at 0 from Lemma 5.1.

LEMMA 10.4. *Given a bounded connected open set O and $w \in O$ there exists a function $\mathcal{G}$ on $O \setminus \{w\}$ satisfying*

(10.31)
(*a*) $\mathcal{G}$ *is* $\mathcal{A} = \nabla f$*-harmonic in* O' *whenever* O' *is open with* $\bar{O}' \subset O \setminus \{w\}$.
(*b*) $\mathcal{G}$ *has boundary value* 0 *on* ∂O *in the* $W^{1,p}$ *Sobolev sense.*
(*c*) *If* $F(x) = G(x-w)$, *for* $x \in \mathbb{R}^n \setminus \{w\}$, *then* $\mathcal{G}(x) \leq F(x)$ *whenever* $x \in O$.
(*d*) $\displaystyle\int_O \langle \nabla f(\nabla \mathcal{G}), \nabla \theta \rangle \, dx = \theta(w)$ *whenever* $\theta \in C_0^\infty(O)$.
(*e*) $\zeta = F - \mathcal{G}$ *extends to a locally Hölder continuous function in* O *and if* O' *is an open set with* $\bar{O}' \subset O$, *then* $\min_{\partial O'} \zeta \leq \zeta(x) \leq \max_{\partial O'} \zeta$ *for* x *in* O'.
(*f*) *There exists* $c \geq 1$ *and* $\delta \in (0,1)$, *depending only on* α, p, n, *and* Λ *in Theorem* A *such that* $|\nabla \zeta(x)| \leq c|x-w|^{\delta-1}$ *for* $x \in B(w, d(w, \partial O)/c)$.
(*g*) $\mathcal{G}$ *is the unique function satisfying* (*a*) $-$ (*d*).

PROOF. We note from Lemma 4.3 that if $B(w, 2/m) \subset O$ and ψ_m is the $\mathcal{A} = \nabla f$-capacitary function for $\bar{B}(w, 1/m)$ with corresponding measure μ_m, then

$$\mu_m(\bar{B}(w,1/m))^{-1/(p-1)} \psi_m \leq c|x-w|^{(p-n)/(p-1)} \tag{10.32}$$

for $x \in \mathbb{R}^n \setminus \bar{B}(w, 2/m)$ where c depends only on the data. Also as in Lemma 5.1 we deduce that the sequence,

$$\left\{\mu_m(\bar{B}(w,1/m))^{-1/(p-1)} \psi_m\right\} \text{ converges to } F \quad \text{as } m \to \infty$$
$$\text{uniformly on compact subsets of } \mathbb{R}^n \setminus \{w\}$$

where F is as in (10.31) (*c*). To construct $\mathcal{G}$, let $\{\tilde{\psi}_m\}_{m \geq 1}$ be a sequence of continuous $\mathcal{A}$-supersolution functions in O with $\tilde{\psi}_m \equiv 1$ on $\bar{B}(w, 1/m)$, while $\tilde{\psi}_m$ is $\mathcal{A}$-harmonic in $O \setminus \bar{B}(w, 1/m)$ with boundary value 0 on ∂O in the $W^{1,p}$ Sobolev sense. Let $\tilde{\mu}_m$ denote the measure corresponding to $\tilde{\psi}_m$. Then from the definition of $\mathcal{A}$-harmonic capacity we see that

$$\tilde{\mu}_m(B(w,1/m)) \geq \mu_m((B(w,1/m))$$

so from (10.32) we have

$$\begin{aligned} \tilde{\mu}_m(\bar{B}(w,1/m))^{-1/(p-1)} \tilde{\psi}_m(x) &\leq \tilde{c} \mu_m(\bar{B}(w,1/m))^{-1/(p-1)} \psi_m(x) \\ &\leq \tilde{c}^2 |x-w|^{(p-n)/(p-1)} \end{aligned} \tag{10.33}$$

for $x \in O \setminus \bar{B}(w, 2/m)$.

Now from (10.33) and the basic estimates in chapter 3 we see that a subsequence of $\{\tilde{\mu}_m(\bar{B}(w,1/m))^{-1/(p-1)} \tilde{\psi}_m(x)\}_{m \geq 1}$ and the corresponding sequence of gradients, converges uniformly on compact subsets of $\mathbb{R}^n \setminus \{w\}$ to an $\mathcal{A}$-harmonic function in $O \setminus \{w\}$ and its gradient which we now denote by $\mathcal{G}, \nabla \mathcal{G}$. Clearly, $\mathcal{G}$ satisfies (10.31) (*a*), (*b*). Also, since

$$\mu_m(\bar{B}(w,1/m))^{-1/(p-1)} \psi_m \to F \quad \text{as} \quad m \to \infty$$

we see from (10.33) that (10.31) (*c*) is true.

From (10.33) and Lemma 3.2 it follows that

$$|\nabla \tilde{\psi}_m(x)| \leq \hat{c}\frac{\tilde{\psi}_m(x)}{|x-w|} \leq \hat{c}^2 \tilde{\mu}(\bar{B}(w,1/m)^{1/(p-1)}\,|x-w|^{(1-n)/(p-1)} \tag{10.34}$$

for $x \in O \setminus \bar{B}(w,4/m)$. From (10.34) we conclude for fixed $q < n(p-1)/(n-1)$ and $m \geq l$, that the sequence

$$\{(\tilde{\mu}(\bar{B}(w,1/m))^{-1/(p-1)}|\nabla \tilde{\psi}_m|\} \text{ is uniformly bounded in } L^q(O \setminus B(w,4/l)) \text{ independent of } l. \tag{10.35}$$

Now (10.35) and uniform convergence of a subsequence of

$$\tilde{\mu}(\bar{B}(w,1/m))^{-1/(p-1)}\nabla \tilde{\psi}_m$$

on compact subsets of $O \setminus \{w\}$ imply that this subsequence also converges strongly in $L^q(O \setminus \{w\})$ to $\nabla \mathcal{G}$ whenever $q < n(p-1)/(n-1)$. Using this fact and writing out the integral identities involving $\tilde{\psi}_m, \tilde{\mu}_m$, we conclude after taking limits, that (10.31) (d) is also valid.

To prove (10.31) (e) we note from the estimate in remark 7.1 that

$$|\nabla F(x)| \approx \langle \nabla F(x), \frac{w-x}{|w-x|}\rangle \approx |x-w|^{(1-n)/(p-1)} \approx F(x)/|x-w| \tag{10.36}$$

whenever $x \in \mathbb{R}^n \setminus \{w\}$ where constants in the ratios depend only on the data. It follows from (10.36) as in the derivations of (5.7)-(5.8), (6.4)-(6.6), (10.11)-(10.13), that $\zeta = F - \mathcal{G}$ is a weak solution to a locally uniformly elliptic PDE in $O \setminus \{w\}$ of the form,

$$\sum_{i,j=1}^{n} \frac{\partial}{\partial x_i}\left(b_{ij}\frac{\partial \zeta}{\partial x_j}\right) = 0 \tag{10.37}$$

where

$$b_{ij}(x) = \int_0^1 f_{\eta_i\eta_j}(t\nabla F(x) + (1-t)\nabla \mathcal{G}(x))dt \quad \text{for } 1 \leq i,j \leq n.$$

Also if $B(w,2r) \subset O$, then for some $c = c(p,n,\alpha,\Lambda) \geq 1$,

$$c^{-1}|\xi|^2|x-w|^{\frac{(p-2)(1-n)}{p-1}} \leq \sum_{i,j=1}^{n} b_{ij}(x)\xi_i\xi_j \leq c|\xi|^2|x-w|^{\frac{(p-2)(1-n)}{p-1}} \tag{10.38}$$

whenever $x \in B(w,r) \setminus \{w\}$. Comparing boundary values of $\mathcal{G}, F$ we observe from the maximum principle for $\mathcal{A}$-harmonic functions and elliptic regularity theory that it suffices to prove (10.31) (e) when $O' = B(w,r)$. To this end, let

$$m(s) = \min_{\partial B(w,s)} \zeta \quad \text{and} \quad M(s) = \max_{\partial B(w,s)} \zeta \quad \text{when} \quad 0 < s \leq r.$$

Let

$$\xi = \liminf_{s \to 0} m(s) \quad \text{and} \quad \beta = \limsup_{s \to 0} M(s).$$

We claim that

$$m(r) \leq \xi = \beta \leq M(r). \tag{10.39}$$

To establish (10.39), first suppose $\xi > M(r)$. In this case, given $0 < N < \xi - M(r)$, we let

$$\theta(x) = \begin{cases} \min[\max(\zeta(x) - M(r), 0), N] & \text{when } x \in B(w,r), \\ 0 & \text{elsewhere in } O. \end{cases}$$

Then $\theta = N$ in a neighborhood of w and vanishes outside of $B(w,r)$ so approximating θ by smooth functions which are constant in a ball about w and taking a limit we see that θ can be used as a test function in (10.31) (d) for both $\mathcal{G}$ and F. Doing this and using the structure assumptions on f in Theorem A it follows that

$$\text{(10.40)} \quad \begin{aligned} c' \int_{\{M(r)+N<\zeta\}} (|\nabla \mathcal{G}| + |\nabla F|)^{p-2} |\nabla \zeta|^2 dx &\leq \int_O \langle \nabla f(\nabla F) - \nabla f(\nabla \mathcal{G}), \nabla \theta \rangle \, dx \\ &= 0. \end{aligned}$$

From (10.40) we see that $\zeta \leq M(r) + N$ almost everywhere in $B(w,r)$ which contradicts our assumption that $\xi > M(r)$. Thus $\xi \leq M(r)$. Next choose a decreasing sequence $\{r_l\}_{l\geq 1}$ with $r_1 = r/2$ and $\lim_{l\to\infty} m(r_l) = \xi$. Applying the minimum principle for $\mathcal{A}$-harmonic functions in $B(w,r_k) \setminus B(w,r_l)$ for $l > k$ and letting $l \to \infty$ we see that

$$\zeta \geq \min(m(r_k), \xi) =: \xi_k \quad \text{in } B(w, r_k).$$

Now using Harnack's inequality in balls $B(y, s/2)$ whenever $y \in \partial B(w,s)$ and $0 < s < r_k/2$ we deduce that

$$M(s) - \xi_k \leq c'(m(s) - \xi_k).$$

Applying this inequality with $s = r_l$, when $r_l < r_k/2$ and letting first $l \to \infty$ and then $k \to \infty$ we find that

$$\liminf_{s\to 0} M(s) = \xi.$$

Now applying the maximum principle once again in a certain sequence of shells with inner radius tending to zero we conclude that $\xi = \beta$. Finally, if $\xi < m(r)$, let $0 < N < m(r) - \xi$ and set

$$\theta(x) = \begin{cases} \min[\max(m(r) - \zeta(x), 0), N] & \text{whenever } x \in B(w,r), \\ 0 & \text{otherwise in } O. \end{cases}$$

Then $\theta \equiv N$ in a neighborhood of w since $\xi = \beta$. Arguing as in the case $\xi > M(r)$, we arrive at a contradiction. Thus (10.39) is valid. Note from (10.39) and arbitrariness of r with $B(w, 2r) \subset O$ that $M(\cdot)$ is increasing and $m(\cdot)$ decreasing on $(0, r_0)$ if $B(w, 2r_0) \subset O$. Using this fact and arguing as in the derivation of (10.27) we get for some $\delta \in (0,1)$ depending only on the data that

$$\text{(10.41)} \qquad M(t) - m(t) \leq (t/s)^\delta (M(s) - m(s)) \quad \text{for} \quad 0 < t \leq s \leq r_0.$$

It follows from (10.41), (10.37)-(10.38), and elliptic regularity theory that ζ is Hölder continuous in $B(w, r_0)$. This completes the proof of (10.31) (e).

To prove (10.31) (f) we note from (10.31) (e) that

$$0 < \zeta(x) \leq \max_{B(w,d(w,\partial O))} F \quad \text{whenever} \quad x \in B(w, d(w, \partial O)).$$

This note, (10.36), and Lemma 10.8 with $\hat{u}_1 = F$, $\hat{u}_2 = \mathcal{G}$, and $O = O \setminus \{w\}$, imply the existence of $c^* \geq 1$ such that

$$\text{(10.42)} \qquad |\nabla \mathcal{G}(x)| \approx \langle \tfrac{w-x}{|w-x|}, \nabla \mathcal{G}(x) \rangle \approx |x-w|^{(1-n)/(p-1)} \approx \frac{\mathcal{G}(x)}{|x-w|}$$

in $B(w, d(w, \partial O)/c^*)$ where c^* and the constants in the ratio all depend only on p, n, α, Λ. From (10.42) and (3.3) it now follows that

$$\text{(10.43)} \qquad |\nabla b_{ij}(x)| \leq c|x-w|^{\frac{-1-n(p-2)}{p-1}}$$

whenever $x \in B(w, d(w, \partial O)/c^*)$ where (b_{ij}) are as in (10.37). Finally, (10.43), Lemma 10.4 (e), and elliptic regularity theory imply Lemma 10.4 (f). $\square$

Next we prove

LEMMA 10.5. *Let D be a starlike Lipschitz domain with center z and let $\mathcal{G}$ be the $\mathcal{A}$-harmonic Green's function for D with pole at z. if*

$$d^*(x) = \min\{d(x, \partial D), |x - z|\}$$

then there exists $c \geq 1$ depending only on the data such that

$$\text{(10.44)}\quad \begin{aligned} &(\alpha) \quad 0 < |\nabla \mathcal{G}(x)| \leq c \left\langle \frac{z - x}{|z - x|}, \nabla \mathcal{G}(x) \right\rangle \quad \textit{whenever } x \in D \setminus \{z\}. \\ &(\beta) \quad c^{-1} \frac{\mathcal{G}(x)}{d^*(x, \partial D)} \leq |\nabla \mathcal{G}(x)| \leq c \frac{\mathcal{G}(x)}{d^*(x, \partial D)} \quad \textit{for } x \in \bar{D} \setminus \{z\}. \end{aligned}$$

PROOF. Since $\mathcal{A}$-harmonic functions are invariant under translation and dilation and (10.44) is also invariant under translation and dilation, we assume, as we may, that

$$z = 0 \quad \text{and} \quad \operatorname{diam}(D) = 1.$$

Let $F, \mathcal{G}$ be as in Lemma 10.4 with $w = 0, O = D$. Using (10.42), starlikeness of ∂D, the maximum principle for $\mathcal{A}$-harmonic functions, and comparing boundary values we see for some $\tilde{c} \geq 1$ and $\gamma > 1$ near 1, that

$$\frac{\mathcal{G}(x) - \mathcal{G}(\gamma x)}{\gamma - 1} \geq \frac{\mathcal{G}(x)}{\tilde{c}} \quad \text{whenever} \quad x \in D \setminus \{0\}$$

where $\tilde{c}$ depends only on the data. Letting $\gamma \to 1$ and using Lemma 3.2 we obtain

$$\text{(10.45)}\qquad -\tilde{c}\langle \nabla \mathcal{G}(x), x \rangle \geq \mathcal{G}(x) \quad \text{when} \quad x \in D \setminus \{0\}.$$

Let

$$\mathcal{P}(x) = -\langle \nabla \mathcal{G}(x), x \rangle \quad \text{whenever} \quad x \in D \setminus \{0\}.$$

From (10.45), (3.2), and the same argument as in (10.6) we deduce that $\phi = \mathcal{G}_{x_i}, 1 \leq i \leq n$, or $\phi = \mathcal{P}$ are weak solutions in $D \setminus \{0\}$ to

$$\text{(10.46)}\qquad \sum_{i,j=1}^{n} \frac{\partial}{\partial x_i} (\hat{b}_{ij} \phi_{x_j}) = 0$$

where

$$\text{(10.47)}\qquad \hat{b}_{ij}(x) = f_{\eta_i \eta_j}(\nabla \mathcal{G}(x)) \quad \text{whenever} \quad x \in D \setminus \{0\}.$$

We temporarily assume that

$$\text{(10.48)}\qquad \mathcal{R} \in C^\infty(\mathbb{R}^n)$$

where $\mathcal{R}$ is as in Definition 9.2. Then as in (10.17) we deduce that $\mathcal{P}$ and $\mathcal{G}_{x_i}, 1 \leq j \leq n$, have continuous extensions to $\bar{D} \setminus \{0\}$. We also have $d(z, \partial D) \approx \operatorname{diam}(D)$ where constants in the ratio depend only on the starlike Lipschitz constant for D. Using (10.42), Lipschitz starlikeness of D, and (10.45) we find for some $\breve{c} \geq 1$ depending only on the data that

$$\breve{c}\mathcal{P}(x) \geq \pm \mathcal{G}_{x_i}(x) \quad \text{on} \quad \partial D \cup B(0, 1/\breve{c}) \setminus \{0\}$$

when $1 \leq i \leq n$. From this inequality and the boundary maximum principle for the PDE in (10.46), we conclude that (10.44) (α) is valid when (10.48) holds with constants depending only on the data. To prove (10.44) (β) note from (10.42) that this inequality is valid in $B(0, \frac{1}{2}d(0, \partial D)) \setminus \{0\}$. Also from Lemma 3.2 ($\hat{a}$) we deduce that the right-hand inequality in (10.44) (β) holds when $x \in D \setminus B(0, \frac{1}{2}d(0, \partial D))$. Thus we prove only the left-hand inequality in (10.44) (β). To do this we first use (10.44) (α) and (10.46), (10.47) for $\mathcal{P}$, once again, to deduce that Moser iteration can be applied to powers of $\mathcal{P}$ in order to obtain,

$$\max_{B(w,s)} \mathcal{P} \leq c \min_{B(w,s)} \mathcal{P} \quad \text{whenever} \quad B(w, 2s) \subset D \setminus \{0\}. \tag{10.49}$$

If $x \in D \setminus B(0, \frac{1}{2}d(0, \partial D))$, we draw a ray l from 0 through x to a point in ∂D. Let y be the first point on l (starting from x) with $\mathcal{G}(y) = \mathcal{G}(x)/2$. Then from the mean value theorem of elementary calculus there exists $\hat{w}$ on the part of l between x, y with

$$\mathcal{G}(x)/2 = \mathcal{G}(x) - \mathcal{G}(y) \leq |\nabla \mathcal{G}(\hat{w})| \, |y - x|. \tag{10.50}$$

From (9.3) with $v = \mathcal{G}, r = 2d(x, \partial D)$, and $x = a_{2r}(w)$, we deduce the existence of $c \geq 1$ depending only on the data with

$$y, \hat{w} \in B(x, (1 - c^{-1})d(x, \partial \hat{D})). \tag{10.51}$$

Using (10.51), the Harnack inequality in (10.49), and (10.44) (α), it follows for some c', depending only on the data, that

$$|\nabla \mathcal{G}(\hat{w})| \leq c' \, |\nabla \mathcal{G}(x)|$$

and thereupon from (10.50) that

$$\mathcal{G}(x) \leq c \, |\nabla \mathcal{G}(x)| \, d(x, \partial \hat{D}).$$

Thus the left-hand inequality in (10.44) (β) is valid when $x \in D \setminus B(0, \frac{1}{2}d(0, \partial D))$ for c suitably large and the proof of Lemma 10.5 is complete under assumption (10.48).

To complete the proof of Lemma 10.5 we show that (10.48) is unnecessary. For this purpose let $\mathcal{R}_m \in C^\infty(\mathbb{R}^n)$ for $m = 1, 2, \ldots,$ with

$$\| \log \mathcal{R}_m \hat{\|}_{\mathbb{S}^{n-1}} \leq c \| \log \mathcal{R} \hat{\|}_{\mathbb{S}^{n-1}} \tag{10.52}$$

and $\mathcal{R}_m \to \mathcal{R}$ as $m \to \infty$ uniformly on $\mathbb{S}^{n-1}$. Here c depends only on n. Let $D_m, \mathcal{G}_m$ be the corresponding starlike Lipschitz domain and $\mathcal{A}$-harmonic Green's function for D_m with pole at 0. Applying Lemma 10.5 to $\mathcal{G}_m$, using Lemmas 3.2, 9.3, and arguing as in the proof of (10.31) (d) we see that

$$\{\mathcal{G}_m, \nabla \mathcal{G}_m\} \text{ converge to } \{\mathcal{G}, \nabla \mathcal{G}\}$$
$$\text{uniformly on compact subsets of } D \setminus \{0\}.$$

Since the constants in this lemma are independent of m we conclude upon taking limits that Lemma 10.5 also holds for $\mathcal{G}$ without hypothesis (10.48). The proof of Lemma 10.5 is now complete. □

Before proceeding further we note the following consequences of Lemma 10.5.

COROLLARY 10.6. *Let D_1, D_2 be starlike Lipschitz domains with center at z and let $\mathcal{G}_1, \mathcal{G}_2$ be the corresponding $\mathcal{A}$-harmonic Green's functions with pole at z. Suppose $w \in \partial D_1 \cap \partial D_2$, $0 < r \leq |w - z|/100$, and*

$$D_1 \cap B(w, 4r) = D_2 \cap B(w, 4r).$$

Then Lemma 10.3 is valid with $\breve{v}_i = \mathcal{G}_i$ *for* $i = 1, 2$, *without assumption* (10.17). *Moreover,* $c_\star$ *in this lemma and so also constants depend only on the data.*

PROOF. From Lemma 10.5 we see that Lemmas 10.2, 10.3 are valid with $\breve{v}_i = \mathcal{G}_i, i = 1, 2$, under assumption (10.17). Also from Lemma 10.5 we deduce that $c_\star$ in these lemmas for $\mathcal{G}_1, \mathcal{G}_2$, depends only on the data. To show assumption (10.17) is unnecessary in Lemma 10.3 let $\mathcal{R}_1, \mathcal{R}_2$, be the graph functions for D_1, D_2 and let $\mathcal{R}_{1,m}, \mathcal{R}_{2,m}, m = 1, 2, \dots$ be approximating graph functions to $\mathcal{R}_1, \mathcal{R}_2$ satisfying (10.52) with $\mathcal{R}_i$ replacing $\mathcal{R}$ for $i = 1, 2$. Also $\mathcal{R}_{i,m} \to \mathcal{R}_i$, as $m \to \infty$, uniformly on $\mathbb{S}^{n-1}$. Finally we choose this sequence so that

$$\mathcal{R}_{1,m} = \mathcal{R}_{2,m} \quad \text{on} \quad \{\omega \in \mathbb{S}^{n-1} : \mathcal{R}_1(\omega) = \mathcal{R}_2(\omega) \in B(w, 4r)\}.$$

Let $D_{i,m}$ be the corresponding starlike Lipschitz domains with center at z and let $\mathcal{G}_{i,m}$ be the $\mathcal{A}$-harmonic Green's functions for $D_{i,m}$ with pole at z for $i = 1, 2$. Applying Lemma 10.3 to $\mathcal{G}_{1,m}, \mathcal{G}_{2,m}$ in $D_{1,m}$ we see that constants in (10.28) depend only on the data. Using Lemma 10.4 and taking limits as $m \to \infty$, we get (10.28) for $\mathcal{G}_1, \mathcal{G}_2$. □

Next we prove,

LEMMA 10.7. *Let* D *be a starlike Lipschitz domain with center* $z, w \in \partial D$, *and* $0 < r \leq |w - z|/100$. *Given* $p, 1 < p < n$, *suppose that* $\tilde{u}, \tilde{v}$ *are positive* $\mathcal{A}$-*harmonic functions in* $D \cap B(w, 4r)$ *and that* $\tilde{u}, \tilde{v}$ *are continuous in* $B(w, 4r) \setminus D$, *with* $\tilde{u}, \tilde{v} = 0$ *on* $B(w, 4r) \setminus D$. *Then there exists* $\tilde{c}_1, 1 \leq \tilde{c}_1 < \infty$, *depending only on the data such that if* $r_1 = r/\tilde{c}_1$, *then*

$$\frac{\tilde{u}(y)}{\tilde{v}(y)} \leq \tilde{c}_1 \frac{\tilde{u}(a_{r_1}(w))}{\tilde{v}(a_{r_1}(w))} \quad \textit{whenever } y \in D \cap B(w, r_1).$$

PROOF. Let $\tilde{w}$ denote the point on the ray from z to w with $|w - \tilde{w}| = \tilde{r} << r$. To prove Lemma 10.7, we assume as we may that $\tilde{u}(\tilde{w}) = 1 = \tilde{v}(\tilde{w})$, since $\mathcal{A}$-harmonic functions are invariant under multiplication by positive constants. Also from (9.3) we see that $\tilde{c} \geq 1$ can be chosen, depending only on the data so that if $\tilde{r} = r/\tilde{c}$, then

$$\max_{B(w,\tilde{r})} \tilde{u} \approx \tilde{u}(\tilde{w}) = 1 \quad \text{and} \quad \max_{B(w,\tilde{r})} \tilde{v} \approx \tilde{v}(\tilde{w}) = 1. \tag{10.53}$$

Second let $\tilde{r}' = \tilde{r}/c$ and let D_1 denote the interior of the domain obtained from drawing all line segments from points in $\partial D \cap \bar{B}(w, \tilde{r}')$ to $\bar{B}(\tilde{w}, \tilde{r}')$. If $c > 10000$ is large enough (depending on the data), and $\tilde{r}' = \tilde{r}/c$, we deduce as in (9.21) that D_1 is starlike Lipschitz with center at $\tilde{w}$. Also the starlike Lipschitz constant for D_1 can be estimated in terms of the starlike Lipschitz constant for D as in Lemma 9.6. Finally there exists $\hat{c} >> c$, depending only on the data such that if $\hat{r} = r/\hat{c} << \tilde{r}'$, then

$$D \cap B(w, 32\hat{r}) = D_1 \cap B(w, 32\hat{r}). \tag{10.54}$$

Let $\mathcal{G}_1$ be the $\mathcal{A}$-harmonic Green's function for D_1 with pole at $\tilde{w}$. Using Harnack's inequality, the maximum principle for $\mathcal{A}$-harmonic functions, the fact that $D_1 \subset D$, (10.31) (e), and (10.36) we obtain that

$$c \min(\tilde{u}, \tilde{v}) \geq r^{(n-p)/(p-1)} \mathcal{G}_1 \quad \text{in} \quad D_1 \setminus B(\tilde{w}, \tilde{r}'/4). \tag{10.55}$$

Let $\mathcal{R}, \mathcal{R}_1$, denote the graph functions for D, D_1 and set

$$K_i := \left\{\omega = \tfrac{y-\tilde{w}}{|y-\tilde{w}|} : y \in B(w, 2^{4-i}\hat{r})\right\} \quad \text{for} \quad i = 0, 1, 2,$$
$$L := \sup_{K_0} \mathcal{R}_1.$$

Choosing $\hat{c}$ still larger if necessary but with the same dependence we see that in addition to (10.54) we may also assume that

$$\{\mathcal{R}_1(\omega)\,\omega : \omega \in K_0\} \subset \partial D \cap \partial D_1\,. \tag{10.56}$$

From our construction we deduce that there exists c_- (depending on p, n, and the Lipschitz constant for $\mathcal{R}_1$) with

$$\min\{d(K_2, \mathbb{S}^{n-1} \setminus K_1), d(K_1, \mathbb{S}^{n-1} \setminus K_0)\} \geq c_-^{-1} \tag{10.57}$$

Let $0 \leq \vartheta \leq 1$ with $\vartheta \in C_0^\infty(\mathbb{R}^n)$, and $\vartheta \equiv 1$ on K_2 while $\vartheta \equiv 0$ on $\mathbb{S}^{n-1} \setminus K_1$. Moreover, thanks to (10.57), we can choose ϑ so that

$$|\nabla \vartheta| \leq \bar{c}_-^{-1} \text{ where } \bar{c}_- \text{ has the same dependence as } c_-\ . \tag{10.58}$$

Let

$$\log \mathcal{R}_2(\omega) := \begin{cases} \vartheta \log \mathcal{R}_1 + (1-\vartheta) \log(2L) & \text{when} \quad \omega \in K_0, \\ \log(2L) & \text{when} \quad \omega \in \mathbb{S}^{n-1} \setminus K_0. \end{cases}$$

Using (10.58), it is easily shown that

$$\|\log \mathcal{R}_2\hat{\|}_{\mathbb{S}^{n-1}} \leq c\,(\|\log \mathcal{R}\hat{\|}_{\partial K_0} + 1).$$

Let D_2 be the starlike Lipschitz domain with center at $\tilde{w}$ and graph function $\mathcal{R}_2$. Also let $\mathcal{G}_2$ be the $\mathcal{A}$-harmonic Green's function for D_2 with pole at $\tilde{w}$. Then from our construction, the fact that $L \approx \tilde{r}$, Harnack's inequality, and once again (10.31) (e), (10.36), we deduce first that

$$c\,r^{(n-p)/(p-1)}\mathcal{G}_2 \geq 1 \quad \text{on} \quad \{\tilde{w} + t\mathcal{R}_1(\omega)\,\omega : \omega \in K_0 \setminus K_1, 0 \leq t \leq 1\}$$

and second from (10.53), (10.56), and the maximum principle for $\mathcal{A}$-harmonic functions that

$$\max(u, v) \leq c\,r^{(n-p)/(p-1)}\mathcal{G}_2 \text{ in } \{\tilde{w} + t\mathcal{R}_1(\omega)\,\omega : \omega \in K_0, 0 < t \leq 1\}. \tag{10.59}$$

Now from (10.56), (10.54), and the definition of $\mathcal{R}_2$, it follows that

$$D \cap B(w, 4\hat{r}) = D_1 \cap B(w, 4\hat{r}) = D_2 \cap B(w, 4\hat{r}). \tag{10.60}$$

Using this display, (10.59), Corollary 10.6 with r replaced by $\hat{r}$, and once again Harnack's inequality we deduce the validity of Lemma 10.7 since r_1 can be chosen so that

$$\frac{\mathcal{G}_1(a_{r_1}(w))}{\mathcal{G}_2(a_{r_1}(w))} \approx 1 \approx \tilde{u}(a_{r_1}(w)) \approx \tilde{v}(a_{r_1}(w)). \qquad \square$$

In order to finish the proof of our boundary Harnack inequalities we need a lemma whose proof requires only Lemmas 3.1-3.2.

LEMMA 10.8. *Let $O \subset \mathbb{R}^n$ be an open set, p fixed, $1 < p < n$ and $\mathcal{A} \in M_p(\alpha)$. Also, suppose that $\hat{v}_1, \hat{v}_2$ are non-negative $\mathcal{A}$-harmonic functions in O.*

Let $\tilde{a} \geq 1$, $y \in O$, $\eta \in \mathbb{S}^{n-1}$, and assume that

$$\frac{1}{\tilde{a}}\frac{\hat{v}_1(y)}{d(y, \partial O)} \leq \langle \nabla \hat{v}_1(y), \eta \rangle \leq |\nabla \hat{v}_1(y)| \leq \tilde{a}\frac{\hat{v}_1(y)}{d(y, \partial O)}.$$

Let $\tilde\epsilon^{-1} = (c\tilde a)^{(1+\tilde\theta)/\tilde\theta}$ *where* $\tilde\theta$ *is as in* (3.3) *of Lemma* 3.2. *If*

$$(1-\tilde\epsilon)\hat L \le \frac{\hat v_2}{\hat v_1} \le (1+\tilde\epsilon)\hat L \quad \textit{in } B(y, \tfrac{1}{100}d(y,\partial O))$$

for some $\hat L, 0 < \hat L < \infty$, *then for* $c = c(p,n,\alpha)$ *suitably large,*

$$\frac{1}{c\,\tilde a}\,\frac{\hat v_2(y)}{d(y,\partial O)} \le \langle \nabla \hat v_2(y), \eta\rangle \le |\nabla \hat v_2(y)| \le c\,\tilde a \frac{\hat v_2(y)}{d(y,\partial O)}.$$

For the proof of similar lemmas see Lemma 3.18 in [**LLN**] and Lemma 5.4 in [**LN1**].

Using Lemma 10.8 and Lemma 10.7 we prove

LEMMA 10.9. *Let* $D, D_1, \mathcal G_1, \tilde u, \tilde v, w, \tilde w, \tilde r, \tilde r', \hat r, c_1, r_1$ *be as in Lemma* 10.7. *Then there exists* $c_2, 1 \le c_1 < c_2 < \infty$, *and* $\theta \in (0,1)$, *depending only on the data, such that if* $r_2 = r/c_2$, *then*

$$\left|\frac{\tilde u(y)}{\tilde v(y)} - \frac{\tilde u(x)}{\tilde v(x)}\right| \le c_2 \left(\frac{|x-y|}{r}\right)^{\theta} \frac{\tilde u(y)}{\tilde v(y)}$$

whenever $x, y \in D \cap B(w, r_2)$.

PROOF. To prove Lemma 10.9 we may assume, as is easily shown using Lemma 10.7. that

$$\tilde u(a_{r_1}(w)) = \tilde v(a_{r_1}(w)) = 1 \quad \text{and} \quad \tilde u = \frac{\mathcal G_1}{\mathcal G_1(a_{r_1}(w))} \text{ in } D\cap B(w,r_1). \tag{10.61}$$

We also temporarily assume that

$$\mathcal R \in C^\infty(\mathbb S^{n-1}). \tag{10.62}$$

From Lemma 10.7 we see that

$$c_1^{-1} \le \frac{\tilde u(y)}{\tilde v(y)} \le c_1 \quad \text{in} \quad D\cap B(w,r_1) \tag{10.63}$$

where $c_1 \ge 1$ depends only on the data. Hence if $\bar u = 2c_1\tilde u$, then

$$\tilde v \le \bar u/2 \le c_1^2 \tilde v \text{ in } D\cap B(w,r_1). \tag{10.64}$$

Let $\{u(\cdot,t)\}$, $0 \le t \le 1$, be the sequence of $\mathcal A$-harmonic functions in $D\cap B(w,r_1/2)$ with continuous boundary values,

$$u(y,t) = t\bar u(y) + (1-t)\tilde v(y), \quad \text{for } 0 \le t \le 1.$$

Existence of $u(\cdot,t), t\in(0,1)$, is proved in [**HKM**]. Checking boundary values and using the maximum principle for $\mathcal A$-harmonic functions, as well as (10.63)-(10.64), we find for some $\tilde c$, depending only on the data, that

$$\frac{u(\cdot,t_1)}{\tilde c} \le \frac{u(\cdot,t_2)-u(\cdot,t_1)}{t_2-t_1} \le \tilde c\, u(\cdot,t_1) \tag{10.65}$$

on $D\cap B(w,r_1/2)$ whenever $0 \le t_1 < t_2 \le 1$.

Let $\epsilon_0 = \tilde\epsilon$ where $\tilde\epsilon$ is as in Lemma 10.8. From (10.65) we find the existence of $\epsilon_0', 0 < \epsilon_0' \le \epsilon_0$, with the same dependence as ϵ_0, such that if $|t_2 - t_1| \le \epsilon_0'$, then

$$1 - \epsilon_0/2 \le \frac{u(\cdot,t_2)}{u(\cdot,t_1)} \le 1 + \epsilon_0/2 \quad \text{in} \quad D\cap B(w,r_1/2).$$

Let $\xi_1 = 0 < \xi_2 < \ldots < \xi_l = 1$ and consider $[0,1]$ as divided into $\{[\xi_k,\xi_{k+1}]\}, 1 \le k \le l-1$. We assume that all of these intervals have a length of $\epsilon_0'/2$ with the possible exception of the interval containing $\xi_l = 1$ which is of length $\leq \epsilon_0'/2$. Using Lemma 10.5 with $\mathcal{G} = \mathcal{G}_1, z = \tilde{w}$, the fact that $u(\cdot,\xi_1) = \bar{u} = 2c_1\tilde{u}$, (10.61), and (10.60), we see that Lemma 10.8 can be applied with $\hat{v}_1 = u(\cdot,\xi_1)$ and $\hat{v}_2 = u(\cdot,t)$, for $0 < t \leq \xi_2$. Doing this we find, for some $c_- \geq 1$ depending only on the data, that

$$c_-^{-1}\frac{u(y,t)}{d(y,\partial D)} \leq |\nabla u(y,t)| \leq c_-\langle \frac{\tilde{w}-y}{|y-\tilde{w}|}, \nabla u(y,t)\rangle \leq c_-^2\frac{u(y,t)}{d(y,\partial D)}, \tag{10.66}$$

whenever $y \in D \cap B(w, r_1/4)$. From (10.66) and (10.65) we see as in (10.37) that if for fixed $t \in [0,\xi_2]$ we define

$$U(\cdot,\tau) = \frac{u(\cdot,t) - u(\cdot,\tau)}{t-\tau} \quad \text{on } D\cap B(w,r_1/4)$$

for $0 \leq \tau < t \leq \xi_2$, then $\mathcal{L}_*^\tau U(\cdot,\tau) = 0$ weakly in $D \cap B(w,r_1/4)$ and $U(\cdot,\tau)$ has continuous boundary value 0 on $\partial D \cap B(w,r_1/4)$, where

$$\mathcal{L}_*^\tau U(x,\tau) = \sum_{i,j=1}^n \frac{\partial}{\partial x_i}\left(b_{ij}^*(x,\tau)\frac{\partial U(x,\tau)}{\partial x_j}\right) = 0 \tag{10.67}$$

and

$$b_{ij}^*(x,\tau) = \int_0^1 f_{\eta_i\eta_j}(s\nabla u(x,t) + (1-s)\nabla u(x,\tau))\,ds \quad \text{for } 1 \leq i,j \leq n. \tag{10.68}$$

Also for some $c \geq 1$, depending only on the data,

$$\sum_{i,j=1}^n b_{ij}^*(x,\tau)\xi_i\xi_j \approx |\xi|^2\left(|\nabla u(x,t)| + |\nabla u(x,\tau)|\right)^{p-2} \tag{10.69}$$

whenever $x \in D\cap B(w,r_1/4)$. Moreover from (10.65), $0 \leq U(\cdot,\tau) \approx u(\cdot,\tau)$ so from (9.2), (9.3), elliptic PDE theory, and Ascoli's theorem it follows that a subsequence of $U(\cdot,\tau)$ converges uniformly as $\tau \to t$ to $V(\cdot)$ with continuous boundary value 0 on $\partial D \cap B(w,r_1/8)$ satisfying $\mathcal{L}_*^t V(\cdot) = 0$ weakly in $D\cap B(w,r_1/8)$. Since $u(\cdot,t)$ is also a weak solution to this PDE we can now apply (10.22) of Lemma 10.2 with $\breve{v}_1 = u(\cdot,t), r = r_1/32$, $h_2 = u(\cdot,t)$, $h_1 = V(\cdot)$, $\hat{x} = \tilde{w}$, to conclude for some $c \geq 100, \theta \in [0,1]$, depending only on the data that if $s_1 = r_1/c$, and $t \in [\xi_1,\xi_2]$. then

$$\left|\frac{V(y)}{u(y,t)} - \frac{V(x)}{u(x,t)}\right| \leq c\left(\frac{|x-y|}{r}\right)^\theta \frac{V(y)}{u(y,t)} \leq c^2\left(\frac{|x-y|}{r}\right)^\theta \tag{10.70}$$

whenever $x,y \in D\cap B(w,s_1)$. To avoid confusion we now write $V(\cdot,t)$ for $V(\cdot)$.

From (10.65) we see for fixed $x \in D\cap B(w,r_1/2)$ that $t \to u(x,t)$ is Lipschitz on $[0,1]$. Hence $u_t(x,t)$ exists for almost every $t\in[0,1]$ and is absolutely continuous on $[0,1]$. Choose a countable dense sequence (x_ν) of $D\cap B(w,r_1/2)$ and a set $W\subset[0,1]$ such that $\mathcal{H}^1([0,1]\setminus W) = 0$ and $u_t(x_\nu,t)$ exists for all $t\in W$ and $\nu = 1,2,\dots$ Then clearly

$$u_t(x_\nu,t) = V(x_\nu,t) \quad \text{for } t\in W \text{ and } \nu = 1,2,\dots$$

Using this equality in (10.70) we deduce that

(10.71)

$$\left|\log\left(\frac{u(x_m,\xi_2)}{u(x_m,\xi_1)}\right) - \log\left(\frac{u(x_k,\xi_2)}{u(x_k,\xi_1)}\right)\right| \leq \int_{\xi_1}^{\xi_2}\left|\frac{u_\tau(x_m,\tau)}{u(x_m,\tau)} - \frac{u_\tau(x_k,\tau)}{u(x_k,\tau)}\right|d\tau \leq c\left(\frac{|x_m-x_k|}{r}\right)^\theta$$

whenever $x_m, x_k \in (x_\nu)$ and $x_m, x_k \in D \cap B(w, s_1)$. As (x_ν) is a dense sequence in $D \cap B(w, s_1)$ it follows from continuity that (10.71) is valid with $x_m = x, x_k = y$, whenever $x, y \in B(w, s_1)$. This validity and (10.64) imply for $x, y \in D \cap B(w, s_1)$. that

$$\left| \frac{u(y,\xi_2)}{u(y,\xi_1)} - \frac{u(x,\xi_2)}{u(x,\xi_1)} \right| \leq c' \left(\frac{|x-y|}{r} \right)^{\theta} \frac{u(y,\xi_2)}{u(y,\xi_1)} \tag{10.72}$$

Using (10.72) we see there exists $\tilde{c}_1' \geq 1$, depending only on the data so that if $s_1' = s_1/\tilde{c}_1'$, then for fixed $y \in D \cap B(w, s_1')$, we have

$$(1 - \epsilon_0/4)\frac{u(y,\xi_2)}{u(y,\xi_1)} \leq \frac{u(x,\xi_2)}{u(x,\xi_1)} \leq (1 + \epsilon_0/4)\frac{u(y,\xi_2)}{u(y,\xi_1)} \quad \text{whenever } x \in D \cap B(w, s_1').$$

If $t \in [\xi_2, \xi_3]$, it follows from our choice of ϵ_0', s_1', that if $x \in D \cap B(w, s_1')$, then

$$\frac{u(x,t)}{u(x,\xi_1)} = \frac{u(x,t)}{u(x,\xi_2)} \cdot \frac{u(x,\xi_2)}{u(x,\xi_1)} \leq (1+\epsilon_0/2)(1+\epsilon_0/4)\frac{u(y,\xi_2)}{u(y,\xi_1)} < (1+\epsilon_0)\frac{u(y,\xi_2)}{u(y,\xi_1)}.$$

for $\epsilon_0 > 0$ small enough. Similarly

$$(1 - \epsilon_0)\frac{u(y,\xi_2)}{u(y,\xi_1)} \leq \frac{u(x,t)}{u(x,\xi_1)} \quad \text{whenever } x \in D \cap B(w, s_1').$$

From these inequalities we obtain that Lemma 10.8 can be applied with $\hat{v}_1 = u(\cdot, \xi_1), \hat{v}_2 = u(\cdot, t)$ and $L = \frac{u(y,\xi_2)}{u(y,\xi_1)}$ in $D \cap B(w, s_1')$. We get that (10.66) holds with constants depending only on the data in $D \cap B(w, s_1'/2)$ whenever $t \in [\xi_2, \xi_3]$. We then can argue as above to eventually conclude that (10.71), (10.72) are valid with ξ_1, ξ_2 replaced by ξ_2, ξ_3 in $D \cap B(w, s_2), s_2 << s_1'$.

We now continue by induction, as in the proof of (4.24) - (4.27) in Theorem 2 of [**LN1**]. We eventually obtain (see [**LN1**, Lemma 4.28]) that (10.66) holds for some constant depending only on the data in $D \cap B(w, s_l)$ whenever $t \in [\xi_{l-1}, \xi_l]$. Also r/s_l depends only on the data. Using this fact and arguing as above we finally obtain for some $c \geq 100$ depending only on the data that if $s_l' = s_l/c$, then (10.71), (10.72) are valid with ξ_1, ξ_2 replaced by ξ_{l-1}, ξ_l in $D \cap B(w, s_l'), s_l' < s_l < s_{l-1}' < \cdots < s_1' < s_1$. Since l also depends only on the data it follows from the l inequalities obtained and the triangle inequality that Lemma 10.9 is true under assumption (10.62).

Finally, we show the assumption (10.62) is unnecessary. Let $\{\mathcal{R}_k\}_{k \geq 1}$ be a sequence of $C^\infty(\mathbb{S}^{n-1})$ functions with

$$\mathcal{R} \leq \mathcal{R}_k \quad \text{and} \quad \|\mathcal{R}_k\hat{\|} \leq c \|\mathcal{R}\hat{\|}$$

satisfying $\mathcal{R}_k \to \mathcal{R}$ uniformly on compact subsets of $\mathbb{S}^{n-1}$. Let $D_k, k = 1, 2, \ldots$ be Lipschitz starlike domains with center z and graph function $\mathcal{R}_k$. Extend $\tilde{u}, \tilde{v}$ to continuous functions on $B(w, 4r) \cap D_k$ by putting $\tilde{u} = \tilde{v} = 0$ in $B(w, 4r) \cap (D_k \setminus D)$.

Let $\tilde{u}_k, \tilde{v}_k$ be $\mathcal{A}$-harmonic functions in $D_k \cap B(w, 3r)$ with $\tilde{u}_k = \tilde{u}, \tilde{v}_k = \tilde{v}$ on $\partial(D_k \cap B(w, 3r))$. Using Lemma 3.2, we see that

$$\{\tilde{u}_k\}_{k \geq 1} \text{ and } \{\tilde{v}_k\}_{k \geq 1} \text{ converge uniformly to } \tilde{u} \text{ and } \tilde{v} \text{ on } B(w, 3r).$$

Moreover, $\{\nabla \tilde{u}_k\}_{k \geq 1}$ and $\{\nabla \tilde{v}_k\}_{k \geq 1}$ converge uniformly to $\nabla \tilde{u}$ and $\nabla \tilde{v}$ on compact subsets of $D \cap B(w, 3r)$. Also, $\tilde{u}_k, \tilde{v}_k$ satisfy the hypotheses of Lemma 10.9 with $3r, D_k$, replacing $4r, D$. We apply this lemma to $\tilde{u}_k, \tilde{v}_k$. Since the constants in this inequality depend only on the data, we can then take limits to get Lemma 10.9 for $\tilde{u}, \tilde{v}$. □

As a corollary to Lemma 10.9 we note that the proof outlined above gives

COROLLARY 10.10. *Let $D, \tilde{w}, \mathcal{A}, \tilde{v}, w, r, r_2$ be as in Lemma 10.9. Then*

$$c^{-1}\frac{\tilde{v}(y)}{d(y,\partial D)} \leq |\nabla \tilde{v}(y)| \leq c\,\langle \frac{\tilde{w}-y}{|\tilde{w}-y|}, \nabla \tilde{v}(y)\rangle \leq c^2 \frac{\tilde{v}(y)}{d(y,\partial D)} \tag{10.73}$$

whenever $y \in D \cap B(w, r_2)$.

PROOF. As noted in the proof of Lemma 10.9 , an induction type argument eventually gives (10.73) for $u(\cdot, \xi_l) = \tilde{v}$ in the smooth case. Taking limits as previously, we then get Corollary 10.10 in general. □

Finally in this chapter we use our results on starlike Lipschitz domains to prove

LEMMA 10.11. *Let $\hat{D}$ be a Lipschitz domain and suppose that*

$$\hat{D} \cap B(w,4r) = \{y = (y', y_n) \in \mathbb{R}^n : y_n > \phi(y')\} \cap B(w,4r)$$

where ϕ is Lipschitz on $\mathbb{R}^{n-1}$, $w \in \partial\hat{D}$, and $r > 0$. Given $p, 1 < p < n$, suppose that $\tilde{u}, \tilde{v}$ are positive $\mathcal{A} = \nabla f$-harmonic functions in $\hat{D} \cap B(w, 4r)$ and that $\tilde{u}, \tilde{v}$ are continuous in $B(w,4r) \setminus \hat{D}$, with $\tilde{u}, \tilde{v} = 0$ on $B(w,4r) \setminus \hat{D}$. Then there exists $c_3, c_4, 1 \leq c_3, c_4 < \infty$, depending only on p, n, α, Λ, and $\|\phi\hat{\|}$ such that if $r_3 = r/c_3$, then

$$\left|\frac{\tilde{u}(y)}{\tilde{v}(y)} - \frac{\tilde{u}(x)}{\tilde{v}(x)}\right| \leq c_3 \left(\frac{|x-y|}{r}\right)^{\theta} \frac{\tilde{u}(y)}{\tilde{v}(y)} \tag{10.74}$$

and if $u^* = \tilde{u}$ or $\tilde{v}$, then

$$c_4^{-1}\frac{u^*(y)}{d(y,\partial\hat{D})} \leq |\nabla u^*(y)| \leq c_4\,\langle e_n, \nabla u^*(y)\rangle \leq c_4^2 \frac{u^*(y)}{d(y,\partial\hat{D})}, \tag{10.75}$$

whenever $x, y \in D \cap B(w, r_3)$.

PROOF. Let $\bar{w} = w + \frac{r}{4}e_n$. As in Lemma 9.6 we observe that if $\bar{c}$ is large enough (depending on p, n and the Lipschitz norm of ϕ), then the domain $D \subset \hat{D} \cap B(w,4r)$ obtained from drawing all open line segments from points in $\partial\hat{D} \cap B(w, r/\bar{c})$ to points in $B(\bar{w}, r/\bar{c})$, is starlike Lipschitz with center $\bar{w}$ and starlike Lipschitz constant bounded above by $c(\||\nabla\phi|\|_\infty + 1)$, where c depends only on n. From this observation, we conclude that (10.74) follows from Lemma 10.9 while (10.75) follows from this observation, Corollary 10.10, and basic geometry. □

CHAPTER 11

Weak convergence of certain measures on $\mathbb{S}^{n-1}$

In this chapter, we use the results in chapters 9 and 10 to fill in some of the details outlined in chapter 8, regarding the pullback of a certain measure under the Gauss map on the boundary of a convex domain. To begin, suppose that E is a compact convex set with 0 in the interior of E and let u be the $\mathcal{A} = \nabla f$-capacitary function for E. Let $\tilde{\mathcal{A}} = \nabla \tilde{f}$, where $\tilde{f}(\eta) = f(-\eta), \eta \in \mathbb{R}^n \setminus \{0\}$. From convexity of E, we see that ∂E is Lipschitz so Corollary 10.10 can be applied to $1-u$ with $\mathcal{A} = \nabla f$ replaced by $\tilde{\mathcal{A}} = \nabla \tilde{f}$. More specifically, from this corollary and basic geometry we see as in Lemma 9.6 that if $w \in \partial E$ and $0 < r_4 \leq r_3$ is small enough, depending only on the data (i.e., the Lipschitz constant for E, p, n, α in Definition 2.1 and Λ in (2.8)) that there is a starlike Lipschitz domain, say $\tilde{\Omega} \subset \mathbb{R}^n \setminus E$ with center at $z \in \mathbb{R}^n \setminus E, |w-z| \approx r_4 \approx d(z, E)$, and

$$\tilde{\Omega} \cap B(w, r_4) = (\mathbb{R}^n \setminus E) \cap B(w, r_4).$$

Moreover, the starlike Lipschitz constant for $\tilde{\Omega}$ can be estimated in terms of the Lipschitz constant for E as in Lemma 9.6. Using these facts and Corollary 10.10 we see that $v = 1-u$ satisfies (9.6) with f replaced by $\tilde{f}$ and D by $\tilde{\Omega}$. Thus Lemma 9.5 and Proposition 9.7 hold for v. It follows that for $\mathcal{H}^{n-1}$-almost every $y \in \partial E$,

$$\lim_{\substack{x \to y \\ x \in \Gamma(y)}} \nabla u(x) = \nabla u(y) \text{ exists} \tag{11.1}$$

and

$$\frac{\nabla u(y)}{|\nabla u(y)|} \text{ is the unit inner normal to } E. \tag{11.2}$$

Here $\Gamma(y)$ is a non-tangential approach region $\subset \mathbb{R}^n \setminus E$ defined below (9.5). Also, if $\Delta(w, r) = \partial E \cap B(w, r)$, and τ is the measure corresponding to $1-u$ as in (9.4), then τ is absolutely continuous with respect to $\mathcal{H}^{n-1}$ on ∂E and

$$d\tau(y) = p \frac{f(\nabla u(y))}{|\nabla u(y)|} d\mathcal{H}^{n-1} \quad \text{for } \mathcal{H}^{n-1}\text{-a.e. } y \in \partial E.$$

Using the above facts and Lemma 9.4 we observe for $0 < r \leq r_4$ that

$$pr^{p-n} \int_{\Delta(w,r)} \frac{f(\nabla u)}{|\nabla u|} d\mathcal{H}^{n-1} = r^{p-n} \tau(\Delta(w, r)) \approx (1 - u(a_{2r}(w)))^{p-1} \tag{11.3}$$

where proportionality constants depend only on the data. Finally, from (11.3), (9.7), and (9.39)-(9.40), and Hölder's inequality we have for $0 < r \leq r_4$,

(11.4)

$$(a)\quad \int_{\Delta(w,r)} \left(\frac{f(\nabla u)}{|\nabla u|}\right)^t d\mathcal{H}^{n-1} \leq c_* r^{(n-1)(1-t)} \left(\int_{\Delta(w,r)} \frac{f(\nabla u)}{|\nabla u|} d\mathcal{H}^{n-1}\right)^t$$

$$(b)\quad \int_{\Delta(w,r)} \mathcal{N}_r(|\nabla u|)^{t(p-1)} d\mathcal{H}^{n-1} \leq c_* r^{(n-1)(1-t)} \left(\int_{\Delta(w,r)} \frac{f(\nabla u)}{|\nabla u|} d\mathcal{H}^{n-1}\right)^t$$

for some $t > p/(p-1)$ and c_* depending only on the data. We note that the non-tangential maximal function $\mathcal{N}_r(\cdot)$ was defined above Lemma 9.5.

Let $\mathbf{g}_E(x) = \mathbf{g} : \partial E \to \mathbb{S}^{n-1}$ be defined by

$$\mathbf{g}_E(x) = -\frac{\nabla u(x)}{|\nabla u(x)|}$$

which is well-defined on a set $\Theta \subset \partial E$ with $\mathcal{H}^{n-1}(\partial E \setminus \Theta) = 0$. From (11.1) and (11.2) we see that if $F \subset \mathbb{S}^{n-1}$ is a Borel set, then $\mathbf{g}^{-1}(F)$ is $\mathcal{H}^{n-1}$ measurable. Define a measure $\mu(\cdot) = \mu_{E,f}(\cdot)$ on $\mathbb{S}^{n-1}$ by

$$\mu(F) := \int_{\Theta \cap \mathbf{g}^{-1}(F)} f(\nabla u)\, d\mathcal{H}^{n-1} \text{ whenever } F \subset \mathbb{S}^{n-1} \text{ is Borel set.} \tag{11.5}$$

Next suppose that $\{E_m\}_{m\geq 1}$ is a sequence of compact convex sets with non-empty interiors which converge to E in the sense of Hausdorff distance. That is, $d_{\mathcal{H}}(E_m, E) \to 0$ as $m \to \infty$ where $d_{\mathcal{H}}$ was defined at the beginning of chapter 2. Let u_m be the corresponding $\mathcal{A} = \nabla f$-capacitary function for E_m and $m = 1, 2, \ldots$ Then for m large enough say $m \geq m_0$ we see that (11.1)-(11.4) hold with u, E replaced by u_m, E_m, when $m \geq m_0$ with constants depending only on the data for E. Let μ_m be the measure on $\mathbb{S}^{n-1}$ defined as in (11.5) with u, E replaced by u_m, E_m. From (11.3), (11.4), we see for some $q > p$ that

$$\int_{\partial E} |\nabla u|^q d\mathcal{H}^{n-1} + \int_{\partial E_m} |\nabla u_m|^q d\mathcal{H}^{n-1} \leq T < \infty \tag{11.6}$$

for $m \geq m_0$ where T depends on the data and the number of balls of radius r_4 needed to cover ∂E. From p-homogeneity of f, (11.6), and Hölder's inequality we see that each of the above measures has finite total mass $\leq \hat{T}$ where $\hat{T}$ has the same dependence as T above. We prove

PROPOSITION 11.1. *Let $\{\mu_m\}_{m\geq 1}$ and μ be measures corresponding to $\{E_m\}_{m\geq 1}$ and E as in* (11.5). *Then*

$$\mu_m \rightharpoonup \mu \quad \textit{weakly as } m \to \infty.$$

Armed with our work in chapters 9 and 10, we could follow [**CNSXYZ**, section 4] which in turn was inspired by the argument in [**J**, section 3]. However, this approach would require that we first prove some preliminary results that were available in the p-harmonic setting. Thus, we give another argument which makes use of the major ideas in [**J**] but which for us was considerably more straight forward. To this end, we first need the following lemma (see [**J**, Lemma 3.3]).

LEMMA 11.2 ([**J**, Lemma 3.3]). *For any $\epsilon > 0$ there exists a positive integer $m_1 = m_1(\epsilon) > m_0$, and a finite collection of disjoint closed balls $\bar{B}(x_j, r_j)$, $1 \leq j \leq N$, with $r_j \leq \epsilon$, $x_j \in \partial E$, and*

$$\mathcal{H}^{n-1}(\partial E \setminus \bigcup_{j=1}^{N} \bar{B}(x_j, r_j)) \leq \epsilon.$$

Moreover, for every $j \in \{1, \ldots, N\}$ and $m \geq m_1$, there exists a rotation and translation, say M_j of $\mathbb{R}^n$, for which $M_j(x_j) = 0$,

$$M_j(E \cap B(x_j, r_j/\epsilon)) = \{(x', x_n) : x_n > \phi(x')\} \cap B(0, r_j/\epsilon),$$
$$M_j(E_m \cap B(x_j, r_j/\epsilon)) = \{(x', x_n) : x_n > \phi_m(x')\} \cap B(0, r_j/\epsilon),$$

where ϕ and ϕ_m are Lipschitz functions on $\mathbb{R}^{n-1}$ satisfying

$$\|\nabla\phi\|_\infty + \|\nabla\phi_m\|_\infty \leq \epsilon. \tag{11.7}$$

PROOF OF PROPOSITION 11.1. Let

$$\Phi := \partial E \setminus \bigcup_{j=1}^{N} \bar{B}(x_j, r_j) \quad \text{and} \quad \Phi_m := \partial E_m \setminus \bigcup_{j=1}^{N} \bar{B}(x_j, r_j) \quad \text{for } m \geq m_1.$$

Let ρ_1, ρ_2, denote respectively the radius of the largest ball contained in the interior of E, and the smallest ball containing E, both with center at the origin. We note that the radial projections from E, E_m, onto $B(0, \rho_1/2)$ are bilipschitz mappings whose bilipschitz constants can be estimated independently of m ($m \geq m_1$) and in terms of $\rho_2/\rho_1, n$. Using this fact and comparing the projections of Φ, Φ_m onto $\mathbb{S}^{n-1}$ we see from Lemma 11.2 that

$$\mathcal{H}^{n-1}(\Phi_m) \leq \kappa\epsilon \quad \text{for } m \geq m_2 \geq m_1, \tag{11.8}$$

where κ is a positive constant independent of m, depending only on the ratio of ρ_2 to ρ_1 and n.

Next for fixed $j, 1 \leq j \leq N$, and M_j as in Lemma 11.2 we let

$$\hat{u}_m(x) := u_m(M_j^{-1}x) \quad \text{and} \quad \hat{u}(x) := u(M_j^{-1}x)$$

when $x \in \mathbb{R}^n$. We also put

$$\hat{E}_m := M_j(E_m) \quad \text{and} \quad \hat{E} := M_j(E).$$

We note that $\hat{u}$ and $\hat{u}_m$ are $\hat{\mathcal{A}} = \nabla\hat{f}$-harmonic in $\mathbb{R}^n \setminus \hat{E}$ and $\mathbb{R}^n \setminus \hat{E}_m$ where $\hat{f}$ satisfies the same structure and smoothness conditions as f. Let

$$C = \{(x', x_n) \in \mathbb{R}^n : |x'| \leq r_j, -r_j < x_n < r_j\}.$$

We also put

$$D' := C \cap (\mathbb{R}^n \setminus \hat{E}) \quad \text{and} \quad D'_m := C \cap (\mathbb{R}^n \setminus \hat{E}_m) \quad \text{for } m \geq m_1.$$

Given $\eta > 0$ small, we claim that if $\epsilon > 0$ is small enough, then

$$|\nabla\hat{u}_m(x', x_n)| \leq c(\epsilon)(x_n - \phi_m(x'))^{-\eta} \tag{11.9}$$

when $x \in B(0, 2r_j) \setminus \hat{E}_m$ and where $c(\epsilon)$ depends on ϵ and the data for E, u, but is independent of m and j for m large enough, say $m \geq m_3 \geq m_2$. The proof of this claim will be given after we use it to prove Proposition 11.1.

We next let $e = \frac{r_j}{2} e_n$ and using the Gauss-Green Theorem in certain approximating domains to D'_m, (11.1)-(11.4) to take limits we deduce as in (9.8)-(9.12) that

$$L_m = (p-1) \int_{\partial D'_m \cap \partial \hat{E}_m} \langle e - x, \nabla \hat{u}_m \rangle \frac{\hat{f}(\nabla \hat{u}_m)}{|\nabla \hat{u}_m|} d\mathcal{H}^{n-1} = J_m + K_m \tag{11.10}$$

where

$$J_m = (n-p) \int_{D'_m} \hat{f}(\nabla \hat{u}_m) dx$$

and

$$K_m = \int_{\partial D'_m \setminus \partial E_m} \langle e - x, \nu_m \rangle \hat{f}(\nabla \hat{u}_m) d\mathcal{H}^{n-1} + \int_{\partial D'_m \setminus \partial E_m} \langle x - e, \nabla \hat{u}_m \rangle \, \langle \nabla \hat{f}(\nabla \hat{u}_m), \nu_m \rangle \, d\mathcal{H}^{n-1}.$$

From Lemmas 3.1 and 3.2 as well as uniqueness of the $\hat{\mathcal{A}}$-capacitary function for a compact convex set with interior, we see that $\{\hat{u}_m\}_{m \geq 1}$ converges uniformly to $\hat{u}$ in $\mathbb{R}^n$ while $\{\nabla \hat{u}_m\}_{m \geq 1}$ converges uniformly to $\nabla \hat{u}$ on compact subsets of $\mathbb{R}^n \setminus E$. Using these facts and claim (11.9) we find from uniform integrability type estimates that

$$\lim_{m \to \infty} J_m = (n-p) \int_{D'} \hat{f}(\nabla \hat{u}) dx \tag{11.11}$$

and

$$\begin{aligned} \lim_{m \to \infty} K_m &= \int_{\partial D' \setminus \partial E} \langle e - x, \nu \rangle \hat{f}(\nabla \hat{u}) d\mathcal{H}^{n-1} \\ &+ \int_{\partial D' \setminus \partial E} \langle x - e, \nabla \hat{u} \rangle \, \langle \nabla \hat{f}(\nabla \hat{u}), \nu \rangle \, d\mathcal{H}^{n-1}. \end{aligned} \tag{11.12}$$

where ν is the outer unit normal to D'. Now (11.10) also holds with D'_m replaced by D' and $\hat{u}_m$ by $\hat{u}$ Using this fact and (11.11)-(11.12) we conclude that

(11.13)

$$\lim_{m \to \infty} \int_{\partial D'_m \cap \partial \hat{E}_m} \langle e - x, \nabla \hat{u}_m \rangle \frac{\hat{f}(\nabla \hat{u}_m)}{|\nabla \hat{u}_m|} d\mathcal{H}^{n-1} = \int_{\partial D' \cap \partial \hat{E}} \langle e - x, \nabla \hat{u} \rangle \frac{\hat{f}(\nabla \hat{u})}{|\nabla \hat{u}|} d\mathcal{H}^{n-1}.$$

Next we note that we can compute the inner normal to $\partial \hat{E}_m$ in two ways: $(\hat{a})$ using (11.1)-(11.2) with $\hat{u}, \hat{E}$, replaced by $\hat{u}_m, \hat{E}_m$ or $(\hat{b})$ using (11.7) of Lemma 11.2 and calculus. Doing this and using the resulting computation in (11.13), we obtain

(11.14)

$$\begin{aligned} &\limsup_{m \to \infty} \left| \int_{\partial D'_m \cap \partial \hat{E}_m} \hat{f}(\nabla \hat{u}_m) d\mathcal{H}^{n-1} - \int_{\partial D' \cap \partial \hat{E}} \hat{f}(\nabla \hat{u}) d\mathcal{H}^{n-1} \right| \\ &\quad \leq c^* \epsilon \left[\limsup_{m \to \infty} \int_{\partial D'_m \cap \partial \hat{E}_m} \hat{f}(\nabla \hat{u}_m) d\mathcal{H}^{n-1} + \int_{\partial D' \cap \partial \hat{E}} \hat{f}(\nabla \hat{u}) d\mathcal{H}^{n-1} \right] \end{aligned}$$

where c^* depends only on the data for E, u. As noted earlier, both integrals on the right-hand side of (11.14) are $\leq \hat{T} < \infty$. We assume as we may that $c^* \epsilon \leq 1/4$.

Then from (11.14) we easily deduce that

$$
\begin{aligned}
&(\alpha)\ \limsup_{m\to\infty}\int_{\partial D'_m\cap\partial\hat{E}_m}\hat{f}(\nabla\hat{u}_m)d\mathcal{H}^{n-1}\le 2\int_{\partial D'\cap\partial\hat{E}}\hat{f}(\nabla\hat{u})d\mathcal{H}^{n-1},\\
&(\beta)\ \limsup_{m\to\infty}\left|\int_{\partial D'_m\cap\partial\hat{E}_m}\hat{f}(\nabla\hat{u}_m)d\mathcal{H}^{n-1}-\int_{\partial D'\cap\partial\hat{E}}\hat{f}(\nabla\hat{u})d\mathcal{H}^{n-1}\right|\\
&\qquad\le 3c^*\epsilon\int_{\partial D'\cap\partial\hat{E}}\hat{f}(\nabla\hat{u})d\mathcal{H}^{n-1}.
\end{aligned}\tag{11.15}
$$

Transferring back to our original scenario using M_j, M_j^{-1}, allowing j to vary, and summing from 1 to N we obtain from (11.14), (11.15), that

$$
\limsup_{m\to\infty}\left|\int_{\partial E_m\setminus\Phi_m}f(\nabla u_m)d\mathcal{H}^{n-1}-\int_{\partial E\setminus\Phi}f(\nabla u)d\mathcal{H}^{n-1}\right|\le c^{**}\epsilon\hat{T},
$$

where $\hat{T}$ was defined above Proposition 11.1 and c^{**} has the same dependence as c^*.

To continue the proof of Proposition 11.1 (under the assumption that claim (11.9) is true), let θ be a continuous function on $\mathbb{S}^{n-1}$ and let

$$
\psi(\rho)=\sup\{|\theta(\omega)-\theta(\omega')|:\ \omega,\omega'\in\mathbb{S}^{n-1},\ |\omega-\omega'|\le\rho\}\quad\text{whenever }\rho\in(0,\infty)
$$

be the modulus of continuity of θ. Let $\mathbf{g}_m=-|\nabla\hat{u}_m|^{-1}\nabla\hat{u}_m$ be the Gauss map defined on a set $\Theta_m\subset\partial E_m$ with $\mathcal{H}^{n-1}(\mathbb{S}^{n-1}\setminus\Theta_m)=0$. Define μ_m as in (11.5) relative to $\mathbf{g}_m, E_m, u_m$ and set

$$
\begin{aligned}
\mu_{j,m}(F)&:=\int_{\Theta_m\cap\bar{B}(x_j,r_j)\cap\mathbf{g}_m^{-1}(F)}f(\nabla u_m)d\mathcal{H}^{n-1}\quad\text{for } j=1,\dots,N,\\
\mu_{0,m}(F)&:=\int_{\Theta_m\cap\Phi_m\cap\mathbf{g}_m^{-1}(F)}f(\nabla u_m)d\mathcal{H}^{n-1}
\end{aligned}
$$

whenever $F\subset\mathbb{S}^{n-1}$ is a Borel set. Define μ_j for $0\le j\le N$, similarly with $u_m,\mathbf{g}_m,E_m,\Theta_m$, replaced by $u,\mathbf{g},E,\Theta$. We note that

$$
\mu_m:=\sum_{j=0}^N\mu_{j,m}\quad\text{and}\quad\mu:=\sum_{j=0}^N\mu_j.
$$

Let j be fixed, $1\le j\le N$, and let

$$
x,x'\in\partial E\cap\Theta\cap B(x_j,r_j)\quad\text{and}\quad y,y'\in\partial E_m\cap\Theta_m\cap B(x_j,r_j).
$$

Then from Lemma 11.2 we see for $m\ge m_1$ that

$$
|\mathbf{g}(x)-\mathbf{g}(x')|+|\mathbf{g}_m(y)-\mathbf{g}_m(y')|+|\mathbf{g}(x)-\mathbf{g}_m(y)|\le c\epsilon \tag{11.16}
$$

where $c=c(n)$. From (11.16) and (11.15) we deduce that

$$
\begin{aligned}
&\limsup_{m\to\infty}\left|\int\theta d\mu_{j,m}-\int\theta d\mu_j\right|\\
&\le\|\theta\|_\infty\limsup_{m\to\infty}|\mu_{j,m}(\mathbb{S}^{n-1})-\mu_j(\mathbb{S}^{n-1})|\\
&\qquad+\psi(c\epsilon)\,[\limsup_{m\to\infty}\mu_{j,m}(\mathbb{S}^{n-1})+\mu_j(\mathbb{S}^{n-1})]\\
&\le 3(c^*\epsilon\|\theta\|_\infty+\psi(c\epsilon))\,\mu_j(\mathbb{S}^{n-1}).
\end{aligned}\tag{11.17}
$$

Also from (11.8), Lemma 11.2, Hölder's inequality, and (11.6), we deduce for $m \geq m_2$ that

$$\begin{aligned}\mu_m(\Phi_m) + \mu(\Phi) &\leq \hat{c}\,[\mathcal{H}^{n-1}(\Phi_m \cup \Phi)]^{1-p/q}\, T^{p/q} \\ &\leq \hat{c}^2 \epsilon^{1-p/q}\, T^{p/q}\end{aligned} \tag{11.18}$$

where $\hat{c}$ depends only on the data for E, u. Summing (11.17) over $1 \leq j \leq N$ we get in view of (11.18), that

$$\limsup_{m\to\infty} \left| \int_{\mathbb{S}^{n-1}} \theta d\mu_m - \int_{\mathbb{S}^{n-1}} \theta d\mu \right| \leq 3(\psi(c\,\epsilon) + c^* \epsilon \|\theta\|_\infty)\hat{T} + \hat{c}^2 \|\theta\|_\infty\, \epsilon^{1-p/q}\, T^{p/q}.$$

Since ϵ can be arbitrarily small and θ is an arbitrary continuous function on $\mathbb{S}^{n-1}$ we conclude that Proposition 11.1 is true under claim (11.9). □

Proof of claim (11.9). Our proof is quite similar to the proof of Lemma 5.28 in [**LN**]. Let $\hat{f}$ be as defined above (11.9) and let

$$K = \{x : x_n/|x| > \cos\hat{\theta}\}$$

be the open spherical cone in $\mathbb{R}^n$ with vertex at the origin, angle opening $\hat{\theta}, \pi/2 < \hat{\theta} < \pi$, and axis parallel to e_n. Let $\bar{u}_l$, be the $\hat{\mathcal{A}} = \nabla\hat{f}$-capacitary function for $\bar{B}(0,l) \setminus K$ and put

$$\bar{v}_l = \frac{1-\bar{u}_l}{1-\bar{u}_l(e_n)} \quad \text{for } l = 1, 2, \ldots.$$

Then $\bar{v}_l$ is $\breve{\mathcal{A}} = \nabla\breve{f}$-harmonic in the complement of $B(0,1) \setminus K$ where $\breve{f}(\eta) = \hat{f}(-\eta), \eta \in \mathbb{R}^n$. Also $\bar{v}_l$ has continuous boundary values with $\bar{v}_l \equiv 0$ on $B(0,l) \setminus K$ and $\bar{v}_l(e_n) = 1$. Using Lemmas 4.1-4.3 and taking limits of a certain subsequence of $\{\bar{v}_l\}_{l\geq 1}$, we see there exists $v \geq 0$, a continuous function on $\mathbb{R}^n$ which is $\breve{\mathcal{A}} = \nabla\breve{f}$-harmonic in K with $v \equiv 0$ on $\mathbb{R}^n \setminus K$ and $v(e_n) = 1$.

We assert that

$$v(tx) = v(te_n)\, v(x) \quad \text{whenever } x \in \mathbb{R}^n \text{ and } t \in (0,\infty). \tag{11.19}$$

To see this, we note that $v(tx), x \in \mathbb{R}^n$ is also $\breve{\mathcal{A}}$-harmonic in K as follows from p-homogeneity of $\breve{f}$. Also $v(tx) = 0$ when $x \notin K$. From these observations we see that Lemma 10.11 can be used with $\tilde{u}, \tilde{v}$ replaced by $v(x), v(tx)$, for $x \in K \cap B(0,R)$ when $R \geq 1$. Moreover, from Harnack's inequality we see that the ratio of $v(Re_n)$ to $v(tRe_n)$ is bounded above and below by constants independent of R when $R \geq 1$. Using these facts and letting $R \to \infty$ we deduce from Lemma 10.11 that $v(tx)$ is a constant multiple of $v(x)$ whenever $x \in K$. From this statement and $v(e_n) = 1$, we obtain (11.19). Differentiating (11.19) with respect to t and evaluating at $t = 1$ we see that

$$\langle x, \nabla v(x)\rangle = \langle e_n, \nabla v(e_n)\rangle v(x) \quad \text{whenever} \quad x \in K.$$

If we let $r = |x|, \omega = x/|x|$ in this identity we obtain that

$$r v_r(r\omega) = \langle e_n, \nabla v(e_n)\rangle v(r\omega).$$

Dividing this equality by $rv(r\omega)$ and integrating with respect to r, we find that $v(\rho\omega) = \rho^\gamma v(\omega)$ whenever $\omega \in \mathbb{S}^{n-1}$ where $\gamma = \langle e_n, \nabla v(e_n)\rangle$. Next we assert that

$$\gamma = \gamma(\hat{\theta}) \to 1 \quad \text{as} \quad \hat{\theta} \to \pi/2. \tag{11.20}$$

To verify this assertion, we write $v = v(\cdot, \hat{\theta}), \gamma = \gamma(\hat{\theta})$ for the above v, γ. If $\pi/2 < \hat{\theta}_k \to \pi/2$ as $k \to \infty$ and a subsequence of $\gamma(\hat{\theta}_k) \to \gamma$ as $k \to \infty$. Then, also a certain subsequence of $v(\cdot, \hat{\theta}_k)$ converges uniformly to v an $\breve{\mathcal{A}}$-harmonic function in $\{x : x_n > 0\}$ with $v(x) = 0$ when $x_n \leq 0$. Since x_n is also $\breve{\mathcal{A}}$-harmonic we conclude as above that v is a constant multiple of x_n and thereupon that (11.20) is true.

Now given $\eta > 0$ as in (11.9) it follows from (11.19) and (11.20) that there exists $\hat{\theta}$ such that

$$\pi/2 < \hat{\theta} < \pi \quad \text{with } \gamma(\hat{\theta}) \geq 1 - \eta.$$

With $\hat{\theta}$ is now fixed, we find from Lemma 11.2 that there exists $\epsilon = \epsilon(\hat{\theta}) > 0$ for which the following is true: Given $\hat{x} \in \partial \hat{E}_m \cap B(0, 2r_j)$ we have $B(\hat{x}, r_j) \setminus (K + \hat{x}) \subset \hat{E}_m$. Let v^* be an $\breve{\mathcal{A}}$-harmonic function with $v^* \equiv 1 - \hat{u}_m$ on $\partial[B(\hat{x}, r_j) \setminus (K + \hat{x})]$. Existence of v^* follows as in [**HKM**]. Comparing boundary values and using the maximum principle for $\breve{\mathcal{A}}$-harmonic functions we have $1 - \hat{u}_m \leq v^*$ in $B(\hat{x}, r_j) \setminus (K + \hat{x})$. Since $x \to v(x - \hat{x})$ is $\breve{\mathcal{A}}$-harmonic in $K + \hat{x}$ we conclude from Lemma 10.3, construction of $v, v^* \leq 1$, and our choice of $\hat{\theta}$ that

$$\begin{aligned}
(1 - \hat{u}_m)(x) &\leq v^*(x) \\
&\leq c \frac{v^*(\hat{x} + r_j e_n/2)}{v(r_j e_n/2)} v(x - \hat{x}) \\
&\leq c(|x - \hat{x}|/r_j)^{1-\eta}
\end{aligned} \tag{11.21}$$

in $B(\hat{x}, r_j) \setminus \hat{E}_m$ where c depends only on the data for $m \geq m_2$. From (11.21) with $x - \hat{x}$ a multiple of e_n and (3.2) we get claim (11.9). In view of our earlier remarks, this finishes the proof of Proposition 11.1. □

CHAPTER 12

The Hadamard variational formula for nonlinear capacity

Let E_1 and E_2 be compact convex sets and suppose 0 is in the interior of $E_1 \cap E_2$. Let $u(\cdot, t)$ be the $\mathcal{A} = \nabla f$-capacitary function for $E_1 + tE_2$ when $t \geq 0$. Also let $\mu_{E_1+tE_2}$ be the measure defined in (8.5) relative to $u(\cdot, t)$. In this chapter we prove

PROPOSITION 12.1. *With the above notation let h_1, h_2 be the support functions for E_1, E_2, respectively and let $\mathbf{g}(\cdot, E_1 + tE_2)$ be the Gauss map for $\partial(E_1 + tE_2)$. Then for $t_2 \geq 0$ we have*

(12.1)

$$\frac{d}{dt} \operatorname{Cap}_{\mathcal{A}}(E_1 + tE_2)\bigg|_{t=t_2} = (p-1) \int_{\mathbb{S}^{n-1}} h_2(\mathbf{g}(z, E_1 + t_2E_2))\, d\mu_{E_1+t_2E_2}(z).$$

PROOF. In the proof of (12.1) we first assume for $i = 1, 2$ that

$$\partial E_i \text{ is locally the graph of an infinitely differentiable and strictly convex function on } \mathbb{R}^{n-1}. \tag{12.2}$$

We note from Lemma 3.2 that $u(\cdot, t_i)$, for $i = 1, 2$, has Hölder continuous second partials in $\mathbb{R}^n \setminus (E_1 + t_2E_2)$. Moreover, as in (5.7)-(5.9) we see that if $0 < t_1 < t_2$, then

$$\zeta(x, t_1) = \frac{u(x, t_1) - u(x, t_2)}{t_1 - t_2} \quad \text{whenever } x \in \mathbb{R}^n,$$

is a positive weak solution to

$$\sum_{i,j=1}^{n} \frac{\partial}{\partial x_i}(\bar{d}_{ij} \zeta_{x_j}) = 0 \tag{12.3}$$

in $\mathbb{R}^n \setminus (E_1 + t_2E_2)$ where

$$\bar{d}_{ij}(x) = \int_0^1 f_{\eta_i\eta_j}(s\nabla u(x, t_1) + (1-s)\nabla u(x, t_2))ds.$$

Also,

$$c^{-1}\bar{\sigma}(x)\,|\xi|^2 \leq \sum_{i,j=1}^{n} \bar{d}_{ij}(x)\xi_i\xi_j \leq c\,\bar{\sigma}(x)\,|\xi|^2 \tag{12.4}$$

whenever $\xi \in \mathbb{R}^n \setminus \{0\}$, $x \in \mathbb{R}^n \setminus (E_1 + t_2E_2)$ and

$$\bar{\sigma}(x) \approx (|\nabla u(x, t_2)| + |\nabla u(x, t_1)|)^{p-2}. \tag{12.5}$$

Constant c in (12.4) depends only on p and n and constants in (12.5) depend only on the structure constants for $\mathcal{A}, p,$ and n. Also from Lemmas 3.2, 4.2, and the

Theorem in [**Li**, Theorem 1] mentioned earlier, we see that $\nabla u(\cdot, t_i)$ for $i = 1, 2$, extend to Hölder continuous functions in the closure of $\mathbb{R}^n \setminus (E_1 + t_i E_2)$. More specifically, if

$$t_2/2 \leq t_1 < t_2 \quad \text{and} \quad \rho = 2t_2 \left(\operatorname{diam}(E_1) + \operatorname{diam}(E_2)\right)$$

then there exist $\beta \in (0, 1)$ and $C^\star \geq 1$, independent of t_1, such that for $i = 1, 2$,

$$\begin{aligned} &(a) \quad |\nabla u(x, t_i) - \nabla u(y, t_i)| \leq C^\star |x - y|^\beta, \\ &(b) \quad (C^\star)^{-1} \leq |\nabla u(x, t_i)| \leq C^\star \end{aligned} \tag{12.6}$$

whenever x, y are in the closure of $B(0, \rho) \setminus (E_1 + t_i E_2)$. From (12.6) (b) and the mean value theorem from calculus we see that ζ is bounded on $\partial(E_1 + t_2 E_2)$ by a constant independent of t_1 when $t_2/2 \leq t_1 < t_2$. Using this fact, (4.4) (b), (4.12) (b) and a weak maximum principle type argument we see for $C^\star$ sufficiently large, independent of $t_1 \in [t_2/2, t_2)$ that

$$\zeta \leq C^\star u(\cdot, t_2) \quad \text{on} \quad \mathbb{R}^n \setminus (E_1 + t_2 E_2). \tag{12.7}$$

Also from (12.6) (a), (12.6) (b), and the Lemmas mentioned above we deduce that

$$\nabla u(\cdot, t_1) \to \nabla u(\cdot, t_2) \text{ uniformly on the closure of } \mathbb{R}^n \setminus (E_1 + t_2 E_2) \text{ as } t_1 \to t_2. \tag{12.8}$$

From (12.8) and (12.3)-(12.5) we see that ζ is locally a bounded solution to a divergence form uniformly elliptic PDE with coefficients that have local Hölder, say $\hat{\beta}$-norm independent of $t_1 \in [t_2/2, t_2)$. From these facts, Lemma 3.2, and Caccioppoli type estimates for locally uniformly elliptic PDE, we deduce that if $t_1 \to t_2$ through an increasing sequence, then a subsequence of the functions corresponding to these values converges uniformly on compact subsets of $\mathbb{R}^n \setminus (E_1 + t_2 E_2)$ to a locally Hölder $\hat{\beta}$ continuous function, say $\breve{\zeta}$, with $\breve{\zeta} \leq C^\star u(\cdot, t_1)$. Moreover, this subsequence also converges to $\breve{\zeta}$ locally weakly in $W^{1,2}$ of $\mathbb{R}^n \setminus (E_1 + t_2 E_2)$. Finally,

$$\sum_{i,j=1}^{n} \frac{\partial}{\partial x_i} (f_{\eta_i \eta_j}(\nabla u(x, t_2))\, \breve{\zeta}_{x_j}(x)) = 0 \quad \text{for} \quad x \in \mathbb{R}^n \setminus (E_1 + t_2 E_2) \tag{12.9}$$

locally in the weak sense in $\mathbb{R}^n \setminus (E_1 + t_2 E_2)$.

Next we show that $\breve{\zeta}$ is independent of the choice of sequence. To do this, for $k = 1, 2$ we let

$$x_k(Z) = \nabla h_k(Z) \quad \text{whenever } Z \in \mathbb{S}^{n-1}.$$

We fix $X, Y \in \mathbb{S}^{n-1}$, write x, y for $x_1(X) + t_2 x_2(X), x_1(Y) + t_2 x_2(Y)$ respectively and note that

$$x = \mathbf{g}^{-1}(X, E_1 + t_2 E_2) \in \partial(E_1 + t_2 E_2) \quad \text{and} \quad y = \mathbf{g}^{-1}(Y, E_1 + t_2 E_2) \in \partial(E_1 + t_2 E_2).$$

We consider two cases. First if

$$|x - y| \leq d(x, E_1 + t_1 E_2)/2$$

then from (12.6) (a) and the mean value theorem of calculus we have

$$|\zeta(x) - \zeta(y) - \langle \nabla \zeta(x), x - y \rangle| \leq \hat{C} |x - y|^\beta \tag{12.10}$$

where $\hat{C}$ is independent of t_1. Second, if

$$|x - y| > d(x, E_1 + t_1 E_2)/2$$

then using $u(\cdot, t_1) \equiv 1$ on $\partial(E_1 + t_1 E_2)$ and the same strategy as above we see that

$$
\begin{aligned}
&|\zeta(x) + \langle \nabla u(x_1(X) + t_1 x_2(X), t_1), x_2(X)\rangle| \\
&\quad + |\zeta(y) + \langle \nabla u(x_1(Y) + t_1 x_2(Y), t_1), y_2(Y)\rangle| \\
&\quad \leq \hat{C}|x-y|^\beta.
\end{aligned}
\tag{12.11}
$$

Now $h_1 + t_1 h_2$ is the support function for $E_1 + t_1 E_2$ and so from (6.23) we have

(12.12)

$$
\begin{aligned}
\langle \nabla u(x_1(X) + t_1(x_2(X)), t_1), x_2(X)\rangle &= |\nabla u(x_1(X) + t_1(x_2(X)), t_1)|\langle X, x_2(X)\rangle \\
&= |\nabla u(x, t_1)| h_2(\mathbf{g}(x, E_1 + t_2 E_2)) + \lambda(x) \\
&= III_1 + \lambda(x)
\end{aligned}
$$

where $\lambda(x) \leq \bar{C}|x-y|^\beta$ and $\bar{C}$ is independent of t_1. Similarly,

(12.13)

$$
\begin{aligned}
\langle \nabla u(x_1(Y) + t_1 x_2(Y), t_1), x_2(Y)\rangle &= |\nabla u(y, t_1)| h_2(\mathbf{g}(y, E_1 + t_2 E_2)) + \bar{\lambda}(y) \\
&= III_2 + \bar{\lambda}(y)
\end{aligned}
$$

where $\bar{\lambda}(y)$ satisfies the same inequality as $\lambda(x)$. From (12.11)-(12.13) and the triangle inequality we find that

$$
\begin{aligned}
|\zeta(x, t_1) - \zeta(y, t_1)| &\leq |III_1 - III_2| + \lambda(x) + \bar{\lambda}(y) \\
&\leq \tilde{C}|x-y|^\beta
\end{aligned}
\tag{12.14}
$$

where $\tilde{C}$ is independent of $t_1 \in [t_2/2, t_2)$. From (12.10), (12.14), we deduce that $\zeta(\cdot, t_1)$ is Hölder β continuous on $\partial(E_1 + t_2 E_2)$ with Hölder norm bounded by a constant independent of $t_1 \in [t_2/2, t_2)$.

We note from (12.9) that ζ is the solution to a divergence form uniformly elliptic PDE with Lipschitz coefficients. Using this note and Hölder β continuity of ζ on $\partial(E_1 + tE_2)$ it follows first (see [**GT**, Theorem 8.29]) that ζ is Hölder θ continuous for some $\theta > 0$ independent of t_1 in $[t_2/2, t_2]$ with Hölder norm on the closure of $B(0, \rho) \setminus \partial(E_1 + t_2 E_2)$ bounded by a constant that is also independent of $t_1 \in [t_2/2, t_2)$. Second,

(12.15)

$$
|\nabla \zeta(x)| \leq \bar{C} d(x, \partial(E_1 + t_2 E_2)^{\theta - 1} \quad \text{whenever} \quad x \in B(0, \rho) \setminus \partial(E_1 + t_2 E_2).
$$

Taking limits we conclude from Ascoli's theorem that $\breve{\zeta}$ is the uniform limit in the closure of $B(0, \rho) \setminus (E_1 + t_2 E_2)$ of a certain subsequence of $\zeta(\cdot, t_1)$ and thus is also Hölder θ continuous in the closure of $B(0, \rho) \setminus (E_1 + t_2 E_2)$. Moreover, (12.15) holds for $\breve{\zeta}$. Finally, arguing as in (12.11)-(12.14) we find that

$$
\zeta(x, t_1) \to |\nabla u(x, t_2)| h_2(\mathbf{g}(x, E_1 + t_2 E_2)) \quad \text{as} \quad t_1 \to t_2
\tag{12.16}
$$

whenever $x \in \partial(E_1 + t_2 E_2)$. From (12.16) we see that every convergent subsequence of $\{\zeta(\cdot, t_1)\}$ converges to a weak solution of (12.9) with continuous boundary values

$$
|\nabla u(\cdot, t_2)|\, h_2(\mathbf{g}(\cdot, E_1 + t_2 E_2)) \quad \text{on} \quad \partial(E_1 + t_2 E_2).
$$

From this deduction, (12.7), (12.9), (4.4) (b), (4.12) (b) and a weak maximum principle argument we conclude that

$$
u_t(x, t)|_{t = t_2} = \lim_{t_1 \to t_2} \zeta(x, t_1) = \breve{\zeta}(x)
\tag{12.17}
$$

whenever x is in the closure of $\mathbb{R}^n \setminus (E_1 + t_2 E_2)$. To begin the proof of Proposition 12.1 in the smooth case and when $t_2/2 \le t_1 < t_2$ we write

$$(t_1 - t_2)^{-1}[\mathrm{Cap}_{\mathcal{A}}(E_1 + t_1 E_2) - \mathrm{Cap}_{\mathcal{A}}(E_1 + t_2 E_2)] = T_1 + T_2 \tag{12.18}$$

where

$$T_1 = (t_1 - t_2)^{-1} \int_{\mathbb{R}^n \setminus (E_1 + t_2 E_2)} (f(\nabla u(x, t_1)) - f(\nabla u(x, t_2)))dx$$

and

$$T_2 = (t_1 - t_2)^{-1} \int_{(E_1 + t_2 E_2) \setminus (E_1 + t_1 E_2)} f(\nabla u(x, t_1))dx.$$

We note from (12.16) that

(12.19)

$$\begin{aligned} T_2 &= p^{-1}(t_1 - t_2)^{-1} \int_{(E_1 + t_2 E_2) \setminus (E_1 + t_1 E_2)} \nabla \cdot [(u(x, t_1) - 1)\, \nabla f(\nabla u(x, t_1))]dx \\ &= -p^{-1} \int_{\partial(E_1 + t_2 E_2)} \zeta(x, t_1) |\nabla u(x, t_2)|^{-1} \langle \nabla f(\nabla u(x, t_1)),\, \nabla u(x, t_2)\rangle \, d\mathcal{H}^{n-1} \\ &\to -\int_{\partial(E_1 + t_2 E_2)} h_2(\mathbf{g}(x, E_1 + t_2 E_2)) f(\nabla u(x, t_2))\, d\mathcal{H}^{n-1} \end{aligned}$$

as $t_1 \to t_2$. Next, we claim that

$$\lim_{t_1 \to t_2} T_1 = \int_{\mathbb{R}^n \setminus (E_1 + t_2 E_2)} \langle (\nabla f)(\nabla u(x, t_2)), \nabla u_{t_2}(x)\rangle dx. \tag{12.20}$$

To prove this claim, first observe that

(12.21)

$$\begin{aligned} T_1 &= (t_1 - t_2)^{-1} \int_{\mathbb{R}^n \setminus (E_1 + t_2 E_2)} \int_0^1 \frac{d}{ds} \left[f(s \nabla u(x, t_1) + (1 - s) \nabla u(x, t_2)) \right] ds dx \\ &= \int_{\mathbb{R}^n \setminus (E_1 + t_2 E_2)} \int_0^1 \langle (\nabla f)(s \nabla u(x, t_1) + (1 - s) \nabla u(x, t_2)), \nabla \zeta(x, t_1)\rangle ds dx. \end{aligned}$$

From local weak convergence in $W^{1,2}$ of $\zeta(\cdot, t_1)$ to u_{t_2}, we have

$$\begin{aligned} &\int_K \int_0^1 \langle (\nabla f)(s \nabla u(x, t_1) + (1 - s) \nabla u(x, t_2)), \nabla \zeta(x, t_1)\rangle ds dx \to \\ &\qquad \int_K \langle (\nabla f)(\nabla u(x, t_2)), \nabla u_{t_2}(x)\rangle dx \end{aligned} \tag{12.22}$$

when $t_1 \to t_2$ for each compact $K \subset \mathbb{R}^n \setminus (E_1 + t_2 E_2)$. Thus to prove (12.20) in view of (12.21) it suffices to show for given $\epsilon > 0$ that if

$$\mathcal{K}(x, s, t_1) = |\nabla f(s \nabla u(x, t_1) + (1 - s) \nabla u(x, t_2))| |\nabla \zeta(x, t_1)|$$

for $s \in [0, 1], t_1 \in [t_2/2, t_2), x \in \mathbb{R}^n \setminus (E_1 + t_2 E_2)$, then there exists $\delta > 0$ small and $R > 0$ large such that

$$\int_{\mathbb{R}^n \setminus B(0, R)} \mathcal{K}(x, s, t_1) dx + \int_{\{x:\ d(x, \partial(E_1 + t_2 E_2)) \le \delta\}} \mathcal{K}(x, s, t_1) dx \le \epsilon. \tag{12.23}$$

Indeed, from (4.12), (12.6), (12.7), and Cacccioppoli type estimates for uniformly elliptic PDE in divergence form, we see that if $E_1 + t_2 E_2 \subset B(0, R/2)$, then

$$
\begin{aligned}
\int_{\mathbb{R}^n \setminus B(0,R)} \mathcal{K}(x, s, t_1) dx &\leq \tilde{C} \int_R^\infty r^{(1-n)/(p-1)}\, dr \\
&= (\tfrac{n-p}{p-1}) \tilde{C} R^{(p-n)/(p-1)} \\
&\leq \epsilon/2
\end{aligned}
\tag{12.24}
$$

for $R \geq R_0$ where $R_0, \tilde{C}$, is independent of s, t_1 in the above intervals. Also from (12.15) we see that

$$
\int_{\{x:\ d(x, \partial(E_1 + t_2 E_2)) \leq \delta\}} \mathcal{K}(x, s, t_1) dx \leq \tilde{C} \delta^\theta \leq \epsilon/2
\tag{12.25}
$$

for $\delta \leq \delta_0$ where δ_0 is independent of s, t_1 in the above intervals. From (12.21)-(12.25) we conclude (12.20).

From (12.21) and (12.15) in the closure of $\mathbb{R}^n \setminus (E_1 + t_2 E_2)$, $\mathcal{A} = \nabla f$-harmonicity of $u(\cdot, t_1)$, p-homogeneity of f, and (12.16), (12.17), we deduce that

$$
\lim_{t_1 \to t_2} T_1 = p \int_{\partial(E_1 + t_2 E_2)} h_2(\mathbf{g}(x, E_1 + t_2 E_2)) f(\nabla u(x, t_2)) d\mathcal{H}^{n-1}.
\tag{12.26}
$$

Combining (12.26) and (12.19), we conclude from (12.18) that

$$
\begin{aligned}
&\frac{d}{dt} \operatorname{Cap}_{\mathcal{A}}(E_1 + tE_2) \,|_{t=t_2} \\
&\quad = \lim_{t_1 \to t_2} \frac{\operatorname{Cap}_{\mathcal{A}}(E_1 + t_2 E_2) - \operatorname{Cap}_{\mathcal{A}}(E_1 + t_1 E_2)}{t_2 - t_1} \\
&\quad = (p-1) \int_{\partial(E_1 + t_2 E_2)} h_2(\mathbf{g}(x, E_1 + t_2 E_2)) f(\nabla u(x, t_2))\, d\mathcal{H}^{n-1}.
\end{aligned}
\tag{12.27}
$$

Now, if $0 < s < 1$ and $t = s/(1-s)$, then

$$
\begin{aligned}
\operatorname{Cap}_{\mathcal{A}}(E_1 + tE_2) &= (1-s)^{p-n} [\operatorname{Cap}_{\mathcal{A}}((1-s)E_1 + sE_2)] \\
&= (1-s)^{p-n} \phi(s)^{n-p}.
\end{aligned}
\tag{12.28}
$$

where ϕ is concave on [0,1] thanks to Theorem A so Lipschitz and differentiable off a countable set. From this observation, the chain rule, and (12.27)-(12.28) we see that Proposition 12.1 is valid under assumption (12.2) except for at most a countable set of $t \in [0, \infty)$. Also, from Proposition 11.1 and properties of support functions we observe that

$$
\psi(t) = (p-1) \int_{\partial(E_1 + tE_2)} h_2(\mathbf{g}(x, E_1 + tE_2))\, f(\nabla u(x, t)) d\mathcal{H}^{n-1}
\tag{12.29}
$$

is continuous as a function of t on $[0, \infty]$. From the Lebesgue differentiation theorem we conclude that Proposition 12.1 is valid under assumption (12.2).

We next remove the assumption (12.2). To this end, choose sequences of uniformly bounded convex domains $\{E_1^{(k)}\}_{k \geq 1}$ and $\{E_2^{(k)}\}_{k \geq 1}$ with $E_i \subset E_i^{(k)}$ for $i = 1, 2$ and $k = 1, 2, \ldots$, satisfying (12.2) with ∂E_i replaced by $\partial E_i^{(k)}, i = 1, 2$ and $k = 1, 2, \ldots,$. We also choose these sequences so that $E_i^{(k)}$ converges to E_i in the sense of Hausdorff distance as $k \to \infty$. Let $\psi_k(t)$ denote the function in (12.29) with

E_i, replaced by $E_i^{(k)}$. Given $0 < a < \infty$ we claim there exists $l = l(a), M = M(a)$, such that for $k \geq l$, we have

$$0 < \psi_k(t) \leq M \quad \text{for} \quad t \in [0, a]. \tag{12.30}$$

To verify this assertion fix k, t, let

$$E_0 = E_1^{(k)} + tE_2^{(k)}$$

and let $h_0, \mathbf{g}_0, u_0$ be the support, Gauss map, and $\mathcal{A} = \nabla f$-capacitary functions corresponding to E_0. Applying Proposition 12.1 in this case with E_1, E_2, t, replaced by $E_0, E_0, 0$, and using the fact that

$$\text{Cap}_{\mathcal{A}}((1+t)E_0) = (1+t)^{n-p}\, \text{Cap}_{\mathcal{A}}(E_0)$$

we get

$$\text{Cap}_{\mathcal{A}}(E_0) = \frac{p-1}{n-p} \int_{\partial E_0} h_0(\mathbf{g}_0(x, E_0))\, f(\nabla u_0(x))\, d\mathcal{H}^{n-1}. \tag{12.31}$$

Since E_0 is uniformly bounded and $h_0 \geq \min_{\mathbb{S}^{n-1}} h_1 > 0$, it follows from (12.31) and properties of capacity, support functions, that (12.30) is true. From (12.30), (12.31), Proposition 11.1, Proposition 12.1 in the smooth case, and the Lebesgue dominated convergence theorem we conclude that

$$\begin{aligned}
\text{Cap}_{\mathcal{A}}(E_1 + tE_2) - \text{Cap}_{\mathcal{A}}(E_1) &= \lim_{k\to\infty} [\text{Cap}_{\mathcal{A}}(E_1^{(k)} + tE_2^{(k)}) - \text{Cap}_{\mathcal{A}}(E_1^{(k)})] \\
&= \lim_{k\to\infty} \int_0^t \psi_k(s) ds = \int_0^t \psi(s) ds.
\end{aligned} \tag{12.32}$$

Also ψ is continuous on $[0, \infty)$ by Proposition 11.1 so (12.32) and the Lebesgue differentiation theorem yield Proposition 12.1 without assumption (12.2). □

REMARK 12.2. *Finally, we remark that Proposition* 12.1 *remains valid for* $t_2 > 0$ *if we assume only that* $0 \in E_1$*, rather than* 0 *is in the interior of* E_1 *(so* $\mathcal{H}^n(E_1) = 0$ *is possible but from the definition of* E_2 *we still have* 0 *in the interior of* E_2*). To handle this case we put* $E_1' = E_1 + t_2 E_2$ *and* $E_2' = E_2$*. Then* E_1', E_2' *are compact convex sets and* 0 *is in the interior of* $E_1' \cap E_2'$*. Applying Proposition* 12.1 *with* E_1, E_2 *replaced by* E_1', E_2' *respectively and at* $t_2 = 0$ *we obtain the above generalization of Proposition* 12.1*.*

We also note that if E_1 *has interior points and* $E_2 = B(0,1)$*, then from the Brunn-Minkowski inequality we have for fixed* $p, 1 < p < n$*, and* $t \in (0, 1)$ *that*

$$\text{Cap}_{\mathcal{A}}(E_1 + tE_2)^{1/(n-p)} - \text{Cap}_{\mathcal{A}}(E_1)^{1/(n-p)} \geq t\, \text{Cap}_{\mathcal{A}}(E_2)^{1/(n-p)}.$$

Dividing this inequality by t *and letting* $t \to 0$ *we get from Proposition* 12.1 *and the chain rule that*

$$\mu_{E_1}(\mathbb{S}^{n-1}) \geq c^{-1}\, \text{Cap}_{\mathcal{A}}(E_1)^{\frac{n-1-p}{n-p}} \tag{12.33}$$

where $c \geq 1$ *depends only on the data.*

CHAPTER 13

Proof of Theorem B

For use in proving Theorem B we shall need the following Lemma.

LEMMA 13.1. *Let $\hat{E}$ and $\hat{E}_l, l = 1, 2, \dots,$ be a sequence of uniformly bounded compact convex sets with $\hat{E}_l \to \hat{E}$ in the Hausdorff distance sense as $l \to \infty$. Then*

$$\lim_{l \to \infty} \mathrm{Cap}_{\mathcal{A}}(\hat{E}_l) = \mathrm{Cap}_{\mathcal{A}}(\hat{E}). \tag{13.1}$$

PROOF. Let $\hat{u}_l$ be the capacitary function for $\hat{E}_l$ and let $\hat{\nu}_l$ be the corresponding capacitary measure for $\hat{E}_l, l = 1, 2, \dots$ defined as in (3.6) (i) relative to $1-u_l$. From (4.4) (a) we deduce that

$$\hat{\nu}_l(\hat{E}_l) = \mathrm{Cap}_{\mathcal{A}}(\hat{E}_l) \quad \text{for} \quad l = 1, 2, \dots. \tag{13.2}$$

From Lemmas 3.2, 3.3, (4.4) (b), and Ascoli's theorem we see that a subsequence of $\{(\hat{u}_l), (\nabla \hat{u}_l)\}$ converges uniformly on compact subsets of $\mathbb{R}^n \setminus \hat{E}$ and locally in $W^{1,p}(\mathbb{R}^n)$ to $\{\hat{u}, \nabla \hat{u}\}$ with $\hat{u}$ an $\mathcal{A}$-harmonic function in $\mathbb{R}^n \setminus \hat{E}$. Moreover, both $\hat{u}$ and $\nabla \hat{u}$ are locally in $W^{1,p}(\mathbb{R}^n)$. On the other hand, taking a subsequence of the subsequence we used to get $\{\hat{u}, \nabla \hat{u}\}$ if necessary and using local uniform boundedness in $W^{1,p}$ of this sequence we may also assume that $\{\hat{\nu}_l\}_{l \geq 1}$ converges weakly to a measure $\hat{\nu}$ with support in $\bar{B}(0, \rho)$. Replacing $\tilde{u}$ in (3.6) (i) by $\hat{u}_l$ and taking limits in (3.6) (i) of the above subsequence, we then get

$$\int \langle \mathcal{A}(\nabla \hat{u}), \nabla \phi \rangle dx = \int \phi \, d\hat{\nu} \quad \text{whenever} \quad \phi \in C_0^\infty(\mathbb{R}^n). \tag{13.3}$$

To prove Lemma 13.1 we consider two cases.

Case 1: If $\mathcal{H}^{n-p}(\hat{E}) < \infty$ then $\mathrm{Cap}_p(\hat{E}) = 0$ so $1 - \hat{u}$ extends to a non-negative $\mathcal{A}$-harmonic function in $\mathbb{R}^n$ (see [**HKM**, Chapter 2]) with

$$1 - \hat{u}(x) \to 0 \quad \text{uniformly as} \quad |x| \to \infty.$$

Using the maximum principle it then follows that $\hat{u} \equiv 1$. So from (13.3) we see that $\hat{\nu}(\mathbb{R}^n) = 0$. Since every subsequence of capacitary measures contains a subsequence converging weakly to the 0 measure we conclude from (13.2) that (13.1) is true in this case and the limit is zero.

Case 2: If $\mathcal{H}^{n-p}(\hat{E}) = \infty$ then using Hausdorff convergence of $\hat{E}_l$ to $\hat{E}$ we see that the constants in (3.5) with $\tilde{E}$ replaced by $\hat{E}_l$ can be chosen independent of $\hat{E}_l$ provided $l \geq l_0$ and l_0 is large enough. From this observation we deduce that $\hat{\sigma}, c$, in (3.4) (ii) applied to $1 - \hat{u}_l$ are independent of l for $l \geq l_0$. Thus, $\{\hat{u}_l\}_{l \geq l_0}^{\infty}$ is locally Hölder continuous with uniform constants on $\mathbb{R}^n$. Using this fact we find that $\hat{u} \equiv 1$ on $\hat{E}$. We conclude that $\hat{u}$ is the capacitary function for $\hat{E}$ and $\hat{\nu}$ the corresponding capacitary measure. Since every subsequence of measures has a subsequence converging weakly to the capacitary measure for $\hat{E}$, we deduce from (4.4) (a) for $\hat{\nu}$ and (13.2) that Lemma 13.1 is true in this case also. □

13.1. Proof of existence in Theorem B in the discrete case

Finally we are in a position to start the proof of existence of the measure in Theorem B in the discrete case. Let $c_1, c_2, \dots, c_m$ be positive numbers and $\xi_i \in \mathbb{S}^{n-1}$, for $1 \le i \le m$. Assume that $\xi_i \neq \xi_j, i \neq j$, and let δ_{ξ_i} denote the measure with a point mass at ξ_i. Let μ be a measure on $\mathbb{S}^{n-1}$ with

$$\mu(K) = \sum_{i=1}^{m} c_i \delta_{\xi_i}(K) \quad \text{whenever} \quad K \subset \mathbb{S}^{n-1} \text{ is a Borel set.}$$

We also assume μ satisfies (8.1) (i) and (ii). That is,

$$\sum_{i=1}^{m} c_i \, |\langle \theta, \xi_i \rangle| > 0 \quad \text{for all} \quad \theta \in \mathbb{S}^{n-1} \tag{13.4}$$

and

$$\sum_{i=1}^{m} c_i \, \xi_i = 0. \tag{13.5}$$

For technical reasons and following either [**J**] or [**CNSXYZ**] we also assume

$$\text{either } \mu(\{\xi\}) \text{ or } \mu(\{-\xi\}) = 0 \quad \text{whenever} \quad \xi \in \mathbb{S}^{n-1}. \tag{13.6}$$

This condition will be removed in our proof of the general case of Theorem B. For μ as above and $p \neq n-1$, we show there is a compact convex polyhedron E with 0 in the interior of E and

$$\mu(K) = \int_{\mathbf{g}^{-1}(K)} f(\nabla U) \, d\mathcal{H}^{n-1} \quad \text{whenever} \quad K \subset \mathbb{S}^{n-1} \text{ is a Borel set}$$

where $\mathbf{g}$ is the Gauss map for ∂E and U is the $\mathcal{A}$-capacitary function for $\mathbb{R}^n \setminus E$. Thus, if F_i denotes the face of ∂E with outer normal $\xi_i, 1 \le i \le m$, then $\mathbf{g}(F_i) = \xi_i$ and

$$\mu(\{\xi_i\}) = c_i = \int_{F_i} f(\nabla U) \, d\mathcal{H}^{n-1} \quad \text{for} \quad 1 \le i \le m. \tag{13.7}$$

If $p = n-1$ our results are less precise. For given μ as above we show the existence of E as above with $\text{Cap}_{\mathcal{A}}(E) = 1$ and corresponding capacitary function U such that for some $b \in (0, \infty)$, (13.7) holds with f replaced by bf.

To set up the minimization problem that will eventually lead to (13.7) (as in [**J, CNSXYZ**]) fix $m, (\xi_i)_1^m, (c_i)_1^m$, and let $q = (q_1, \dots, q_m) \in \mathbb{R}^m$ with $q_i \ge 0, 1 \le i \le m$. Let

$$E(q) := \bigcap_{i=1}^{m} \{x : \ \langle x, \xi_i \rangle \le q_i\} \quad \text{and} \quad \Theta := \{E(q) : \ \text{Cap}_{\mathcal{A}}(E(q)) \ge 1\}.$$

We also set

$$\gamma(q) = \sum_{i=1}^{m} c_i \, q_i \quad \text{with} \quad \gamma = \inf\{\gamma(q) : \ E(q) \in \Theta\}.$$

We want to show there exists

$$\breve{q} = (\breve{q}_1, \dots, \breve{q}_m), \ \breve{q}_i > 0 \text{ for } 1 \le i \le m \text{ with } \gamma(\breve{q}) = \gamma \text{ and } \text{Cap}_{\mathcal{A}}(E(\breve{q})) = 1. \tag{13.8}$$

Once (13.8) is proved we can use the same argument as in [**J**, Section 5] or [**CNSXYZ**, Section 6] to get (13.7).

To begin the proof of (13.8) we first note that if $E(q) \in \Theta$, then $E(q)$ is a closed convex set. Also we note from (13.5) that

$$\int_{\mathbb{S}^{n-1}} \langle \tau, \xi \rangle^+ d\mu(\xi) = \int_{\mathbb{S}^{n-1}} \langle \tau, \xi \rangle^- d\mu(\xi) \quad \text{whenever} \quad \tau \in \mathbb{S}^{n-1}$$

where $a^+ = \max(a, 0)$ and $a^- = \max(-a, 0)$. From this note and (13.4) we see that for some $\phi > 0$,

$$\phi < \int_{\mathbb{S}^{n-1}} \langle \tau, \xi \rangle^+ d\mu(\xi) \quad \text{for all} \quad \tau \in \mathbb{S}^{n-1}. \tag{13.9}$$

If $r\tau \in E(q)$, it follows from (13.9) that $r \leq \gamma(q)/\phi$ so

$$E(q) \subset \{x : \ |x| \leq \gamma(q)/\phi\}. \tag{13.10}$$

From (13.10) we deduce the existence of $q^l = (q_1^l, \ldots, q_m^l), q_i^l \geq 0, 1 \leq i \leq m$, such that $E_l = E(q^l)$ for $l = 3, 4, \ldots$, is a sequence of uniformly bounded compact convex sets in Θ, with

$$\hat{q} = \lim_{l \to \infty} q^l \quad \text{and} \quad \lim_{l \to \infty} \gamma(q^l) = \gamma = \gamma(\hat{q}).$$

From finiteness of γ we also may assume (by taking further subsequences if necessary) that $E_l \to E(\hat{q}) = E_1$, a compact convex set containing 0, uniformly in the Hausdorff distance sense. From Lemma 13.1 we observe that

$$\lim_{l \to \infty} \operatorname{Cap}_{\mathcal{A}}(E_l) = \operatorname{Cap}_{\mathcal{A}}(E_1). \tag{13.11}$$

It follows that $\operatorname{Cap}_{\mathcal{A}}(E_1) \geq 1$ and $E_1 \in \Theta$. In fact $\operatorname{Cap}_{\mathcal{A}}(E_1) = 1$, since otherwise we would have $\gamma(\tilde{q}) < \gamma(\hat{q})$ for $\tilde{E} = \tilde{E}(\tilde{q}) \in \Theta$ where for $j \in \{1, 2, \ldots, m\}$,

$$\tilde{q}_j = \frac{\hat{q}_j}{\operatorname{Cap}_{\mathcal{A}}(E_1)^{1/(n-p)}}.$$

Next we consider two cases.

Case 1: First suppose that z is an interior point of E_1. Then $\breve{E} = E_1 - z \in \Theta$, since the distance from 0 to each plane composing the boundary of $\breve{E}$ is positive. Also, $\breve{E}$ has $\mathcal{A}$-capacity 1 and if $\breve{E} = E(\breve{q})$, then from (13.5) we see that $\gamma(\breve{q}) = \gamma$. Thus, (13.8) is valid in this case.

Case 2: If E_1 has empty interior, then from convexity of E_1 and (13.6) we find that E_1 is contained in a $k < n - 1$ dimensional plane and $0 < \mathcal{H}^k(E_1) < \infty$. Moreover we must have $p > n - k$ since otherwise as mentioned in Lemma 13.1 we have $\operatorname{Cap}_{\mathcal{A}}(E_1) = 0$. Also, we may assume 0 is an interior point of E_1 relative to the k-dimensional plane containing E_1 since otherwise we consider $E_1 - z$ for some z having this property and argue as above. In this case from the definition of Θ and (13.4), we see that there exists a subset, say Λ of $\{1, \ldots, m\}$ with $\hat{q}_i = 0$ when $i \in \Lambda$. From (13.6) we deduce that Λ has cardinality at least 3. Also since a point has $\mathcal{A}$-capacity zero we see that $\{1, \ldots, m\} \setminus \Lambda$ contains at least two points. Moreover, since E_1 is a minimizer we observe that if $s \notin \Lambda$, then $\hat{q}_s \neq 0$ and

$$\{x : \ \langle x, \xi_s \rangle = \hat{q}_s\} \cap E_1 \neq \emptyset.$$

Let $a = \frac{1}{4}\min\{\hat{q}_i : i \notin \Lambda\}$ and for small $t > 0$ let

$$\tilde{E}(t) = \bigcap_{i=1}^{m} \{x : \langle x, \xi_i \rangle \le \hat{q}_i + at\}$$
$$E_2 = \bigcap_{i=1}^{m} \{x : \langle x, \xi_i \rangle \le a\}. \tag{13.12}$$

Put

$$E_t = \frac{\tilde{E}(t)}{\mathrm{Cap}_{\mathcal{A}}(\tilde{E}(t))^{1/(n-p)}}. \tag{13.13}$$

We note that, in view of (13.12), $E_t = E(q(t))$ where $q(t) = (q_1(t), \dots, q_m(t))$ and

$$q_j(t) = \frac{\hat{q}_j + at}{\mathrm{Cap}_{\mathcal{A}}(\tilde{E}(t))^{1/(n-p)}} \quad \text{for} \quad 1 \le j \le m. \tag{13.14}$$

From properties of $\mathcal{A}$-capacity, we have $\mathrm{Cap}_{\mathcal{A}}(E_t) = 1$ so $E_t \in \Theta$. To get a contradiction to our assumption that E_1 has empty interior we show that

$$\gamma(q(t)) < \gamma \quad \text{for some small } t > 0. \tag{13.15}$$

To prove (13.15), we first note that $E_1 + tE_2 \subset \tilde{E}(t)$ for $t \in (0, 1)$ so

$$\mathrm{Cap}_{\mathcal{A}}(E_1 + tE_2) \le \mathrm{Cap}_{\mathcal{A}}(\tilde{E}(t)).$$

From this inequality and (13.13), (13.14), we conclude that to prove (13.15) it suffices to show if

$$k(t) = \mathrm{Cap}_{\mathcal{A}}(E_1 + tE_2)^{-1/(n-p)} \sum_{i=1}^{m} c_i(\hat{q}_i + at)$$

then

$$k(t) < \gamma \quad \text{for } t > 0 \text{ near } 0. \tag{13.16}$$

To prove (13.16), we let, as in chapter 12, $u(\cdot, t)$ be the $\mathcal{A} = \nabla f$-capacitary function for $E_1 + tE_2$ and let $\mathbf{g}(\cdot, E_1 + tE_2)$ be the Gauss map for $\partial(E_1 + tE_2)$ while h_1, h_2 are the support functions for E_1, E_2, respectively. Then from Remark 12.2 and Proposition 12.1 we have for $t \in (0, 1)$,

$$\frac{d}{dt}\mathrm{Cap}_{\mathcal{A}}(E_1 + tE_2) = (p-1)\int_{\partial(E_1 + tE_2)} h_2(\mathbf{g}(x, E_1 + tE_2)) f(\nabla u(x, t)) d\mathcal{H}^{n-1}. \tag{13.17}$$

Next we prove

PROPOSITION 13.2.

$$\lim_{\tau \to 0} \int_{\partial(E_1 + \tau E_2)} h_2(\mathbf{g}(x, E_1 + \tau E_2)) f(\nabla u(x, \tau)) d\mathcal{H}^{n-1} = \infty. \tag{13.18}$$

Assuming Proposition 13.2 we get (13.15) and so a contradiction to our assumption that E_1 has empty interior as follows. First observe from (13.17) that

(13.19)

$$(n-p)[\mathrm{Cap}_{\mathcal{A}}(E_1+tE_2)]^{1+1/(n-p)}\frac{d}{dt}k(t)\Big|_{t=\tau} = (n-p)\,\mathrm{Cap}_{\mathcal{A}}(E_1+\tau E_2)\sum_{i=1}^{m} c_i a$$

$$-(p-1)[\sum_{i=1}^{m} c_i(\hat{q}_i+a\tau)]\int_{\partial(E_1+\tau E_2)} h_2(\mathbf{g}(x,E_1+\tau E_2))f(\nabla u(x,\tau))d\mathcal{H}^{n-1}.$$

Now $E_1+\tau E_2\to E_1$ as $\tau\to 0$ in the sense of Hausdorff distance so by Lemma 13.1, we have

$$\lim_{\tau\to 0}\mathrm{Cap}_{\mathcal{A}}(E_1+\tau E_2)=\mathrm{Cap}_{\mathcal{A}}(E_1)=1. \tag{13.20}$$

Clearly, (13.18)-(13.20) imply for some $t_0>0$ small that

$$\frac{d}{dt}k(t)\Big|_{t=\tau}<0 \quad \text{for } \tau\in(0,t_0]. \tag{13.21}$$

Also, from (13.20) we see that

$$\lim_{\tau\to 0}k(\tau)=\gamma.$$

From this observation, the mean value theorem from calculus, and (13.21) we conclude that (13.16) holds so E_1 has interior points. (13.8) now follows from our earlier remarks.

PROOF OF PROPOSITION 13.2. Recall that E_1 is contained in a $k<n-1$ dimensional plane and $n-k<p<n$. We assume as we may that

(13.22)

$$E_1\subset\{x=(x',x''):x'=(x_1,\dots,x_k)\quad\text{and}\quad x''=(x_{k+1},\dots x_n)=(0,\dots,0)\}=\mathbb{R}^k.$$

Indeed, otherwise we can rotate our coordinate system to get (13.22) and corresponding $\hat{\mathcal{A}}$-capacitary functions, say $\bar{u}(\cdot,t)$. Proving Proposition 13.2 for $\bar{u}(\cdot,t)$ and transferring back we obtain Proposition 13.2.

We also note that

$$\bar{B}(0,4a)\cap\mathbb{R}^k\subset E_1\subset\bar{B}(0,\rho) \tag{13.23}$$

which follows from our choice of a and for some ρ large (depending only on the data). We shall need the following lemma.

LEMMA 13.3. *There exists $C_1\geq 1$, such that if $\psi=(p-n+k)/(p-1)$ and $x\in B(0,\rho)$, then*

$$|x''|^{\psi}\leq C_1(1-u(x,t))\quad \textit{whenever}\quad C_1t\leq|x''| \tag{13.24}$$

where $C_1\geq 1$ depends on various quantities but is independent of $x\in B(0,2\rho)$ and t.

PROOF OF LEMMA 13.3. To prove this Lemma we note from Lemma 5.3 in [**LN4**] that there exists an $\tilde{A}(\eta)=-(\nabla f)(-\eta)$-harmonic function $\hat{V}$ on $\mathbb{R}^n\backslash\mathbb{R}^k$ with continuous boundary value 0 on $\mathbb{R}^k$ and $\hat{V}(x)\approx|x''|^{\psi}$ for $x\in\mathbb{R}^n$ where constants depend only on p,n,k and the structure constants for f. For fixed $t\in(0,1)$, let $v=\max(\hat{V}-C_2t,0)$. Then v is $\tilde{A}(\eta)$-harmonic in $\mathbb{R}^n\setminus W$ and continuous on $\mathbb{R}^n$ with $v\equiv 0$ on $W=\{x:\hat{V}(x)\leq C_2t\}$. From the definition of E_1+tE_2 and v we see for C_2 large enough, depending on p, n, k, the structure constants for f, and ρ, that $v = 0$ on $E_1 + tE_2$. Also from (13.23), (4.4) of Lemma 4.2, Harnack's inequality, and the fact that $E_1 + tE_2$ has $\mathcal{A}$-capacity ≥ 1 we find that $C_3(1 - u(\cdot, t)) \geq v$ on $\partial B(0, 2\rho)$ where C_3 has the same dependence as C_2 and $u(x, t)$ is the $\tilde{\mathcal{A}}$-capacitary function for $E_1 + tE_2$. Using the maximum principle for $\tilde{\mathcal{A}}$-harmonic functions it now follows that $v \leq C_4(1 - u(\cdot, t))$ in $B(0, 2\rho)$. From this fact and our knowledge of $\hat{V}$ we get Lemma 13.3. □

To begin the proof of Proposition 13.2 we assume $0 < t \leq \tilde{t}_0$, where $\tilde{t}_0 << a$. We also observe that $E_1 + tE_2$ is a compact convex set with nonempty interior so from Corollary 10.10 and Proposition 9.7 we find that for $\mathcal{H}^{n-1}$ almost every $\hat{x} \in \partial(E_1 + tE_2)$

$$\nabla u(y, t) \to \nabla u(\hat{x}, t) \quad \text{as} \quad y \to \hat{x}$$

non-tangentially in $\mathbb{R}^n \setminus (E_1 + tE_2)$. Moreover, there exists $\tilde{c}$ such that $B(\hat{x}, 4t/\tilde{c}) \cap \partial(E_1 + tE_2)$ is the graph of a Lipschitz function whenever

$$\hat{x} \in B(0, 2a) \cap \partial(E_1 + tE_2) \quad \text{and} \quad 0 < t \leq \tilde{t}_0$$

with Lipschitz constant independent of $\hat{x}, t$. It then follows from (9.5), (9.7)(a), and (9.39)-(9.40) that

(13.25)
$$\begin{aligned} c \int_{B(\hat{x}, t/\tilde{c}) \cap \partial(E_1 + tE_2)} f(\nabla u(\cdot, t)) d\mathcal{H}^{n-1} &\geq (1 - u(w, t))^p t^{n-1-p} \\ &\geq c^{-1} \int_{B(\hat{x}, t/\tilde{c}) \cap \partial(E_1 + tE_2)} f(\nabla u(\cdot, t)) d\mathcal{H}^{n-1} \end{aligned}$$

where c depends only on the data and $w = w(\hat{x}, t)$ denotes a point in $B(\hat{x}, t/\tilde{c}) \cap (\mathbb{R}^n \setminus (E_1 + tE_2))$ whose distance from $\partial(E_1 + tE_2)$ is $\geq t/c^2$. Using Harnack's inequality in a chain of balls of radius $\approx t$ connecting w to a point $x \in B(0, a)$ with $2C_1 t = |x''|$ we deduce from (13.24) of Lemma 13.3 that

$$1 - u(w, t) \geq C^{-1} t^{\psi} \tag{13.26}$$

where C is independent of $t \in (0, 1)$. Using (13.26) in (13.25) we obtain for some $C' \geq 1$, independent of $t, 0 < t \leq \tilde{t}_0$, that

$$C' \int_{B(\hat{x}, t/\tilde{c}) \cap \partial(E_1 + tE_2)} f(\nabla u(\cdot, t)) d\mathcal{H}^{n-1} \geq t^{p(\psi - 1) + n - 1}. \tag{13.27}$$

Now since $\partial(E_1 + tE_2) \cap B(0, 2a)$ projects onto a set containing $B(0, 2a) \cap \mathbb{R}^k$ for $0 < t \leq \tilde{t}_0$, we see there is a disjoint collection of balls $B(\hat{x}, t/\tilde{c})$ for $\hat{x} \in \partial(E_1 + tE_2)$ of cardinality approximately t^{-k} for which (13.27) holds. Since

$$p(\psi - 1) + (n - 1) - k = (k + 1 - n)/(p - 1) < 0$$

we conclude from (13.27) that for some C^* independent of small positive t

(13.28)
$$C^* \int_{\partial(E_1 + tE_2) \cap B(0, 2a)} f(\nabla u(\cdot, t)) d\mathcal{H}^{n-1} \geq t^{(k+1-n)/(p-1)} \to \infty \quad \text{as} \quad t \to 0.$$

Finally note that for $0 < t \leq \tilde{t}_0$,

$$\mathbf{g}(x, E_1 + tE_2) \in \{\xi_i : i \in \Lambda\}$$

for $\mathcal{H}^{n-1}$ almost every $x \in \partial(E_1 + tE_2) \cap B(0, 2a)$ and $h_2(\xi_i) \equiv a$ whenever $\xi_i \in \Lambda$. From this note and (13.28), we obtain first the validity of (13.18) in Proposition 13.2 and thereupon that (13.8) is true. $\square$

Armed with (13.8), we now complete the proof of existence in Theorem B in the discrete case. Given $q^* = (q_1^*, \dots, q_m^*) \in \mathbb{R}^m$ with $q_i^* > 0$ for $1 \le i \le m$, we note from (13.8) that for $\bar{t}_0 > 0$ sufficiently small, as in the remark following (13.15) that $E(q^*(t)) \in \Theta$ for $0 < t \le t_0$, where

$$q^*(t) = \frac{(1-t)\hat{q} + tq^*}{\mathrm{Cap}_{\mathcal{A}}((1-t)E(\hat{q}) + tE(q^*))^{1/(n-p)}}.$$

Also, $\gamma(q^*(t)) \ge \gamma$ for $0 \le t \le \bar{t}_0$ thanks to (13.8). Now as in (13.19), we have for $\tau > 0$ small,

(13.29)

$$\begin{aligned}
&(n-p)[\mathrm{Cap}_{\mathcal{A}}((1-t)E(\hat{q}) + tE(q^*))]^{1+1/(n-p)} \left.\frac{d\gamma(q^*(t))}{dt}\right|_{t=\tau} \\
&\quad = (n-p)\,\mathrm{Cap}_{\mathcal{A}}((1-\tau)E(\hat{q}) + \tau E(q^*)) \sum_{i=1}^m c_i(q_i^* - \hat{q}_i) \\
&\qquad - \left[\sum_{i=1}^m c_i((1-\tau)\hat{q}_i + \tau q_i^*)\right] \frac{d}{dt}\,\mathrm{Cap}_{\mathcal{A}}((1-t)E(\hat{q}) + tE(q^*))|_{t=\tau}\,.
\end{aligned}$$

Moreover, using

$$\mathrm{Cap}_{\mathcal{A}}((1-t)E(\hat{q}) + tE(q^*)) = (1-t)^{n-p}\,\mathrm{Cap}_{\mathcal{A}}(E(\hat{q}) + sE(q^*))$$

where $s = t/(1-t)$, Proposition 12.1, and the chain rule we have with $\hat{h}, h^*$ the support functions for $E(\hat{q}), E(q^*)$, and $u^*(\cdot, s)$ the $\mathcal{A}$-capacitary function for $E(\hat{q}) + sE(q^*)$,

(13.30)

$$\begin{aligned}
&(1-t)^{p+2-n} \frac{d}{dt}\,\mathrm{Cap}_{\mathcal{A}}((1-t)E(\hat{q}) + tE(q^*))|_{t=\tau} \\
&\quad = -(n-p)(1-\tau)\,\mathrm{Cap}_{\mathcal{A}}(E(\hat{q}) + \tau E(q^*)) \\
&\qquad + (p-1) \int_{\partial(E(\hat{q}) + \tau E(q^*))} h^*(\mathbf{g}(\cdot, E(\hat{q}) + \tau E(q^*)))\, f(\nabla u^*(\cdot, \tau)) d\mathcal{H}^{n-1}.
\end{aligned}$$

Using (12.31) with $E_0 = E(\hat{q})$ in (13.30) and letting $t \to 0$, we conclude from (13.29), (13.11), and the mean value theorem in elementary calculus that

(13.31)

$$\begin{aligned}
&0 \le (n-p) \lim_{\tau \to 0} \frac{d\gamma(q^*(\tau))}{d\tau} \\
&\quad = (n-p) \sum_{i=1}^m c_i(q_i^* - \hat{q}_i) - (p-1)\gamma \int_{\partial E(\hat{q})} (h^* - \hat{h})(\mathbf{g}(x, E(\hat{q})))\, f(\nabla u^*(x, 0)) d\mathcal{H}^{n-1} \\
&\quad = (n-p) \sum_{i=1}^m c_i(q_i^* - \hat{q}_i) - (p-1)\gamma \sum_{i=1}^m (q_i^* - \hat{q}_i) \int_{\mathbf{g}^{-1}(\xi_i, E(\hat{q}))} f(\nabla u^*(x, 0)) d\mathcal{H}^{n-1}
\end{aligned}$$

provided q^* is near enough $\hat{q}$. Clearly, the possible choices of $q^* - \hat{q}$ contain an open neighborhood of 0. Thus

$$c_i = \left(\frac{p-1}{n-p}\right)\gamma \int_{\mathbf{g}^{-1}(\xi_i, E(\hat{q}))} f(\nabla u^*(\cdot,0))\, d\mathcal{H}^{n-1} \quad \text{for } 1 \leq i \leq m. \tag{13.32}$$

From (13.32) and p-homogeneity of f we find that if $p \neq n-1$, and $E = \phi E(\hat{q})$ where $\phi^{n-p-1} = (\frac{p-1}{n-p})\gamma$, and U is the $\mathcal{A} = \nabla f$-capacitary function corresponding to E, then (13.7) holds. If $p = n-1$, put $b = (\frac{p-1}{n-p})\gamma$, $E = E(\hat{q})$, and $U = u^*(\cdot, 0)$. This completes the proof of existence in the discrete case when (13.4)-(13.6) hold.

REMARK 13.4. *We note for later use from* (13.32), *the definition of* ϕ, (12.31) *with* $E_0 = E$, *and* (12.33) *that if* r *denotes the radius of a ball with* $\mathcal{A}$*-capacity* 1 *and* h *is the support function for* E *as in* (13.7) *when* $1 < p < n, p \neq n-1$, *or its amended form when* $p = n-1$, *then*

(13.33)
$$\int_{\mathbb{S}^{n-1}} h(\xi) d\mu(\xi) = \frac{n-p}{p-1}\operatorname{Cap}_{\mathcal{A}}(E)$$
$$\begin{cases} = \frac{n-p}{p-1}\operatorname{Cap}_{\mathcal{A}}(E) \leq c\mu(\mathbb{S}^{n-1})^{\frac{n-p}{n-p-1}} & \text{when } 1 < p < n-1, \\ \leq \gamma = b/(n-2) \leq r\mu(\mathbb{S}^{n-1}) & \text{when } p = n-1, \\ = \frac{n-p}{p-1}\operatorname{Cap}_{\mathcal{A}}(E) \leq c(\operatorname{diam}(E))^{n-p} & \text{when } n-1 < p < n. \end{cases}$$

13.2. Existence in Theorem B in the continuous case

Armed with existence in Theorem B in the discrete case, we now consider existence when μ is a finite positive measure on $\mathbb{S}^{n-1}$ satisfying (8.1). We choose a sequence of discrete measures $\{\mu_j\}_{j\geq 1}$ satisfying (13.4)-(13.6) when p is fixed $1 < p < n$ with

$$\mu_j \rightharpoonup \mu \quad \text{weakly as} \quad j \to \infty.$$

If $p \neq n-1$, we let $E_j, j = 1, 2, \ldots$, be a corresponding sequence of compact convex sets with nonempty interiors and $\mathcal{A}$-capacitary functions U_j for which (13.7) holds at support points of μ_j.

If $p = n-1$, we choose E_j and corresponding capacitary function U_j with $\operatorname{Cap}_{\mathcal{A}}(E_j) = 1$ satisfying the amended form of (13.7) for $j = 1, 2, \ldots$ Then $\mu_j = b_j\tilde{\mu}_j$ for some $b_j > 0$, where $\tilde{\mu}_j$ is the measure in (8.5) (b) with u, μ replaced by $U_j, \tilde{\mu}_j$ for $j = 1, 2, \ldots$

From the definition of weak convergence we may assume for some $C \geq 1$ that

$$C^{-1} \leq \mu_j(\mathbb{S}^{n-1}) \leq C \quad \text{for } j = 1, 2, \ldots \tag{13.34}$$

and thanks to (13.4) that for some $\hat{C} \geq 1$,

$$\hat{C}^{-1} \leq \int_{\mathbb{S}^{n-1}} |\langle\theta, \xi\rangle|\, d\mu_j(\xi) \quad \text{whenever} \quad \theta \in \mathbb{S}^{n-1} \text{ and } j = 1, 2, \ldots. \tag{13.35}$$

Following [**J**] or [**CNSXYZ**], we claim that we may also assume

$$E_j \subset \bar{B}(0, \rho) \text{ for } j = 1, 2, \ldots, \text{ and some } \rho < \infty. \tag{13.36}$$

To prove (13.36) we first note from (13.5) that we may translate E_j if necessary so that if $d_j = \operatorname{diam}(E_j)$ then the line segment from $-d_jy_j/2$ to $d_jy_j/2$ is contained in E_j for some $y_j \in \mathbb{S}^{n-1}$ when $j = 1, 2, \ldots$. If h_j denotes the support function for E_j, then from the definition of support function it follows that

$$h_j(\xi) \geq \tfrac{1}{2}|\langle d_j y_j, \xi\rangle| \quad \text{whenever} \quad \xi \in \mathbb{S}^{n-1}. \tag{13.37}$$

Using (13.37), (13.35), (13.33), and (13.34), we deduce that

$$\begin{aligned}
(2\hat{C})^{-1} d_j &\leq \frac{d_j}{2} \int_{\mathbb{S}^{n-1}} |\langle y_j, \xi\rangle| \, d\mu_j(\xi) \leq \int_{\mathbb{S}^{n-1}} h_j(\xi) d\mu_j(\xi) \\
&= \left(\frac{n-p}{p-1}\right) \mathrm{Cap}_{\mathcal{A}}(E_j) \\
&\leq \begin{cases} \tilde{C} & \text{for } 1 < p < n-1, \\ b_j/(n-2) \leq \tilde{C} & \text{for } p = n, \\ c d_j^{n-p} \leq \tilde{C} & \text{for } n-1 < p < n \end{cases}
\end{aligned} \tag{13.38}$$

where $\tilde{C}$ is a positive constant that does not depend on j. Thus, claim (13.36) is true.

From (13.36) we see that a subsequence of $\{E_j\}_{j\geq 1}$ (also denoted $\{E_j\}$) converges to a compact convex set $E \subset \bar{B}(0, \rho)$ in the sense of Hausdorff distance.

We proceed by considering the following two cases.

Case A: E has nonempty interior. In this case, if $p \neq n-1$, it follows from weak convergence of measures in Proposition 11.1 that Theorem B is true.

To handle the possibility that $p = n-1$, and for later use, fix $p, 1 < p < n$, and if $j = 1, 2, \ldots$, let ν_j denote the capacitary measure corresponding to U_j as in (4.4) (a) of Lemma 4.2. Then from Lemma 9.5 and (9.39) of Proposition 9.7 we see that ν_j is absolutely continuous with respect to $\mathcal{H}^{n-1}$ on ∂E_j and

$$\frac{d\nu_j}{d\mathcal{H}^{n-1}}(y) = p \frac{f(\nabla U_j(y))}{|\nabla U_j(y)|} \quad \text{for} \quad \mathcal{H}^{n-1} \text{ almost every } y \in \partial E_j. \tag{13.39}$$

Let $\hat{r} = \sup\{s : B(x, s) \subset E\}$ be the inner radius of E. Since E_j converges to E in the sense of Hausdorff distance it follows that E_j has inner radius at least $\hat{r}/2$ and from (13.36) that $E_j \subset \bar{B}(0, \rho)$ for $j \geq j_0$, provided j_0 is large enough. Using these facts and convexity of E_j, it follows from basic geometry that if $\hat{x} \in \partial E_j$, then after a possible rotation,

$$B(\hat{x}, \hat{r}/100) \cap E_j = \{x = (x', x_n) : x_n > \hat{\phi}(x')\}$$

where $\hat{\phi}$ is a Lipschitz function on $\mathbb{R}^{n-1}$ with $\|\hat{\phi}\|_{\mathbb{R}^{n-1}} \leq c(\rho/\hat{r})$. Moreover, ∂E_j can be covered by at most $c(\frac{\rho}{\hat{r}})^{n-1}$ of radius $\hat{r}/1000$ where c depends only on the data. From these observations, (13.39), as well as the reverse Hölder inequality in (9.7) (a) with $p = q$ and our discovery in Corollary 10.10 that for $j \geq j_0$ sufficiently large j_0, there exists $\breve{C}$ depending only on the data and $\hat{r}, \rho$, such that

$$\begin{aligned}
\mu_j^*(\mathbb{S}^{n-1}) &= \int_{\partial E_j} f(\nabla U_j) d\mathcal{H}^{n-1} \\
&\leq \breve{C}(\hat{r})^{(1-n)/(p-1)} \left(\int_{\partial E_j} p \frac{f(\nabla U_j)}{|\nabla U_j|} d\mathcal{H}^{n-1} \right)^{p/(p-1)} \\
&= \breve{C}(\hat{r})^{(1-n)/(p-1)} \left(\nu_j(\partial E_j)\right)^{p/(p-1)}
\end{aligned} \tag{13.40}$$

where $\mu_j^* = \mu_j$ if $p \neq n-1$ and $\mu_j^* = \tilde{\mu}_j$ if $p = n-1$.

Next as in the discussion following (13.2), we see that if $\mathrm{Cap}_{\mathcal{A}}(E) \neq 0$ then a subsequence of $\{\nu_j\}$ (also denoted $\{\nu_j\}$) satisfies,

$$\lim_{j\to\infty} \nu_j = \nu \quad \text{weakly where } \nu \text{ is the capacitary measure for } E. \tag{13.41}$$

We now finish the proof of Theorem B when E has nonempty interior. From (3.5) we deduce $\mathrm{Cap}_{\mathcal{A}}(E) \neq 0$. Then as in (13.1) and (13.2) we observe that

$$\lim_{j\to\infty} \nu_j(\partial E_j) = \lim_{j\to\infty} \mathrm{Cap}_{\mathcal{A}}(E_j) = \mathrm{Cap}_{\mathcal{A}}(E) = \nu(E) = 1. \tag{13.42}$$

Using (13.42) in (13.40) we conclude that $\{\tilde{\mu}_j\}$ is uniformly bounded for $j \geq j_0$. In view of (13.38) and Proposition 11.1 we can choose subsequences of $\{b_j\}$ and $\{\tilde{\mu}_j\}$ (also denoted $\{b_j\}$ and $\{\tilde{\mu}_j\}$) so that

$$\lim_{j\to\infty} b_j = b,\ 0 < b < \infty, \quad \text{and} \quad \lim_{j\to\infty} \tilde{\mu}_j = \tilde{\mu} \text{ weakly} \tag{13.43}$$

where $\tilde{\mu}$ is the measure in (8.5) (b) with u, μ replaced by $U, \tilde{\mu}$. Also U is the capacitary function for E. Then, using (13.43), we have $\mu = b\,\tilde{\mu}$ so the proof of Theorem B is complete when $1 < p < n$ and E has nonempty interior.

Case B: E has empty interior. In this case, we assume that

$$\mathrm{Cap}_{\mathcal{A}}(E) \neq 0. \tag{13.44}$$

At the end of this section, we show that (13.44) holds. Since sets with finite $\mathcal{H}^{n-p}$ measure have zero $\mathcal{A}$-capacity, it then follows from (13.36) as in the discrete case that there is a k-dimensional plane P with $n - p < k \leq n - 1$ and

$$E \subset P \cap B(0, \rho) \quad \text{with} \quad 0 < \mathcal{H}^k(E) < \infty. \tag{13.45}$$

By considering two cases, $n - p < k < n - 1$ and $k = n - 1$, we show that (13.45) is not possible and this will finish the proof that E has nonempty interior.

Case B1: Suppose that $n - p < k < n - 1$. Translating and rotating E if necessary, we may assume (13.22)-(13.23) hold for some $a > 0$ with E_1 replaced by E and ρ by 2ρ.

Let

$$t_j = d_{\mathcal{H}}(E_j, E) \quad \text{for } j = 1, 2, \ldots.$$

Then for j large enough we can argue as in Lemma 13.3 with t replaced by t_j, and $u(\cdot, t), E_1 + tE_2$, by U_j, E_j. We obtain from the analogue of (13.24) for $j \geq j_0$, that

$$1 - U_j(x) \geq C_1^{-1}\, |x''|^{\psi} \quad \text{for } x = (x', x'') \in B(0, 4\rho) \text{ and } C_1\, t_j \leq |x''| \tag{13.46}$$

where $x' \in \mathbb{R}^k$ and $\psi = (p - n + k)/(p - 1)$. Fix $j \geq j_0$, and given $y \in \partial E_j \cap B(0, a)$, let $T_j(y)$, be a supporting hyperplane to ∂E_j at y. Let $\hat{H}_j$ be the open half space with $\hat{H}_j \cap E_j = \emptyset$ and $\partial \hat{H}_j = T_j(y)$. Let y^* denote the point in $\hat{H}_j$ which lies on the normal line through y with $|y - y^*| = 2C_1 t_j$ where C_1 is as in (13.46). Note that for j sufficiently large it is true that $d(y^*, P) > C_1 t_j$. since otherwise it would follow from the triangle inequality that there exists $z \in B(0, 2a) \cap E$ with $d(z, E_j) > t_j$. Thus (13.46) holds with $x = y^*$. Let ϕ be the $\mathcal{A}$-harmonic function in $\hat{H}_j \cap B(y, 8C_1 t_j) \setminus \bar{B}(y^*, C_1 t_j)$ with continuous boundary values

$$\phi \equiv \begin{cases} 1 - U_j & \text{on } \partial B(y^*, C_1\, t_j), \\ 0 & \text{on } \partial(\hat{H}_j \cap B(y, 8C_1 t_j)). \end{cases}$$

Then from the maximum principle for $\tilde{\mathcal{A}}$-harmonic functions we have $\phi \leq 1 - U_j$ on $\hat{H}_j \cap B(y, 8C_1 t_j) \setminus B(y^*, C_1\, t_j)$. Comparing ϕ to a linear function and using a boundary Harnack inequality from [**LLN**] in $\hat{H}_j \cap B(y, 8C_1 t_j)$ we deduce for some c^* depending only on the data and ρ that

$$(1 - U_j(y^*))/t_j \leq c^*(1 - U_j(\hat{z}))/d(\hat{z}, T_j(y))$$

when $\hat{z} \in \hat{H}_j \cap B(y, C_1 t_j)$. Letting $\hat{z} \to y$ non-tangentially, we conclude from this inequality and (13.46) with $x = y^*$ that

$$t_j^{\psi-1} \leq C^{**}|\nabla U_j(y)| \quad \text{for } \mathcal{H}^{n-1} \text{ almost every } y \in \partial E_j \cap B(0, a) \tag{13.47}$$

and $j \geq j_0$. Here C^{**} has the same dependence as C_1. From (13.47), Theorem B in the discrete case, (13.39) with ν replaced by ν_j, and the structural assumptions on f we see that

$$t_j^{\psi-1}\, \nu_j(\partial E_j \cap B(0, a)) \leq C' \mu_j^*(\mathbf{g}_j(\partial E_j \cap B(0, a))) \tag{13.48}$$

where $\mu_j^* = \mu_j$ when $p \neq n-1$, and $\mu_j^* = \tilde{\mu}_j$ when $p = n-1$ for $1 < p < n$. Here $\mathbf{g}_j$ is the Gauss map for ∂E_j. We note that $\psi - 1 = (1 - n + k)/(p-1) < 0$. Also from (13.41) we deduce

$$\liminf_{j \to \infty} \nu_j(\partial E_j \cap B(0, a)) \geq \nu(\partial E \cap B(0, a/2)) > 0$$

where the last inequality follows from the fact that otherwise U extends to an $\mathcal{A}$-harmonic function in $B(0, a/2)$ which would then imply by Harnack's inequality that $U \equiv 1$. Using this inequality in (13.48) we see that if $p \neq n-1$, then $\mu_j(\mathbb{S}^{n-1}) \to \infty$ in contradiction to (13.34). If $p = n-1$, then from (13.38) we see that

$$\inf\{b_j,\ j = 1, 2, \dots\} > 0$$

since otherwise it would follow that E consists of a single point - a set with zero $\mathcal{A}$-capacity. Using this observation and arguing as above, we get once again that $\mu_j(\mathbb{S}^{n-1}) \to \infty$ which again contradicts (13.34). Thus, we first conclude that E can not be contained in a k-dimensional plane for $n - p < k < n-1$ under the assumption (13.44).

Case B2: Suppose that $k = n-1$. In this case, we continue our proof under the assumption that (13.44) holds. We also assume, as we may, that $P = \{x : x_n = 0\}$ and

$$B(0, 4a) \cap P \subset E \subset B(0, \rho) \cap P. \tag{13.49}$$

Let U be the $\mathcal{A}$-capacitary function for E and as previously, U_j is the $\mathcal{A}$-capacitary function for E_j. Translating E_j slightly upward if necessary we may assume that

$$\lim_{j \to \infty} d_{\mathcal{H}}(E_j, E) = 0 \quad \text{and} \quad E_j \subset \{x : x_n > 0\}.$$

Let $\nabla U_+(x)$ denote the limit (whenever it exists) as $y \to x$ non-tangentially through values with $y_n > 0$. We prove

PROPOSITION 13.5. *There exists $C \geq 1$ such that*

$$C \liminf_{j \to \infty} \int_{\partial E_j} f(\nabla U_j) d\mathcal{H}^{n-1} \geq \int_{\partial E} f(\nabla U_+) d\mathcal{H}^{n-1} - C^2 \mathcal{H}^{n-1}(E). \tag{13.50}$$

PROOF. Given $\epsilon > 0$ choose j_1 so large that $d_{\mathcal{H}}(E_j, E) \leq \epsilon$ for $j \geq j_1$. Comparing boundary values of U, U_j in $B(0, 2\rho) \setminus E_j$ and using Lemmas 3.3, 4.3, 4.4, we deduce the existence of $0 < \alpha \leq 1/2, \hat{C} \geq 1$, such that

$$1 - U \leq \hat{C}(1 - U_j + \epsilon^\alpha) \tag{13.51}$$

for $j \geq j_1$. Next we divide the interior of E into $(n-1)$-dimensional closed Whitney cubes $\{Q_k\}$. Let $\partial' E$ denote the boundary of E considered as a set in P. Then the cubes in $\{Q_k\}$ have disjoint interiors with side length $s(Q_k)$ and the property that considered as sets in P, the distance say $d'(Q_k, \partial' E)$ from Q_k to the boundary of E satisfies

$$10^{-n} s(Q_k) \leq d'(Q_k, \partial' E) \leq 10^n s(Q_k). \tag{13.52}$$

Let $Q \in \{Q_k\}$ with $s(Q) \geq \epsilon^\alpha$ and put $Q_+ = Q \times (0, s(Q))$. Suppose $y = (y_1, \dots, y_n) = y_Q$ is a point in $Q_+ \setminus E_j$ with $d(y, \partial' E) \geq y_n/2 \geq s(Q)/4$.

We consider two possibilities. If $(1-U)(y) \geq 2\hat{C}\epsilon^\alpha$ ($\hat{C}$ as in (13.51)), then from (13.51) we have $1 - U(y) \leq 2\hat{C}(1 - U_j(y))$ and arguing as in (13.25) for U_j, U we get

$$\begin{aligned} \bar{C}^3 \int_{\partial E_j \cap Q_+} f(\nabla U_j) d\mathcal{H}^{n-1} &\geq \bar{C}^2 (1 - U_j)^p(y)\, s(Q)^{n-1-p} \\ &\geq \bar{C}(1 - U)^p(y)\, s(Q)^{n-1-p} \\ &\geq \int_Q f(\nabla U_+) d\mathcal{H}^{n-1}. \end{aligned} \tag{13.53}$$

If $1 - U(y) < 2\hat{C}\epsilon^\alpha$, then since $s(Q) \geq \epsilon^\alpha$, an argument similar to the above gives

$$\int_Q f(\nabla U_+) d\mathcal{H}^{n-1} \leq C_+\, s(Q)^{n-1} \tag{13.54}$$

where C_+ is independent of $j \geq j_2 \geq j_1$ provided j_2 is large enough. Combining (13.53), (13.54) and using (13.52) we find after summing over $Q \in \{Q_k\}$ that for some $\breve{C} \geq 1$, independent of $j \geq j_2$,

$$\breve{C} \int_{\partial E_j} f(\nabla U_j) d\mathcal{H}^{n-1} \geq \int_{\{x \in E:\, d'(x, \partial' E) \geq \breve{C}\epsilon^\alpha\}} f(\nabla U_+) d\mathcal{H}^{n-1} - \breve{C}^2 \mathcal{H}^{n-1}(E). \tag{13.55}$$

Letting first $j \to \infty$ and after that $\epsilon \to 0$ we obtain from (13.55) and the monotone convergence theorem or Fatou's Lemma that (13.50) is true. This finishes the proof of Proposition 13.5. □

Next we prove

PROPOSITION 13.6.

$$\int_E f(\nabla U_+) d\mathcal{H}^{n-1} = \infty. \tag{13.56}$$

We note that Propositions 13.5, 13.6 give a contradiction to our assumption that (13.49) is true, since the total mass of the measures in (13.34) are uniformly bounded and invariant under translation. Hence, we also conclude in this case that E can not be contained in k-dimensional plane when $k = n - 1$ under the assumption (13.44). This finishes the proof of existence in Theorem B under the assumption (13.44).

PROOF OF PROPOSITION 13.6. For the readers benefit we first outline a "simple" proof of (13.56) in Proposition 13.6 for the p-Laplace equation (i.e., when $(f(\eta) = p^{-1}|\eta|^p)$). We use the same notation as in Proposition 13.5. Let $\{Q_k\}$ be a Whitney decomposition of the interior of E considered as a subset of P. Let

$Q \in \{Q_k\}$, and let z be a point in $\partial' E$ with $d'(z, Q) \approx s(Q)$. From convexity of E we see that there is a $(n-2)$-dimensional plane, say P_1 containing z with the property that E is contained in the closure of one of the components of $P \setminus P_1$. Rotating P_1 if necessary we may assume that

$$E \subset \Omega = \{x \in \mathbb{R}^n : \ x_1 - z_1 \leq 0 \ \text{ and } \ x_n = 0\}.$$

We note that Krol in [**Kr**] has constructed a homogeneous $1 - 1/p$ solution, say v', to the p-Laplace equation in $\{(x_1, x_n)\} \setminus \{(x_1, 0) : x_1 \leq 0\}$ which vanishes continuously on $(-\infty, 0] \times \{0\}$. We extend v' continuously to $\mathbb{R}^n$ (also denoted v') by defining this function to be constant in the other $(n-2)$ coordinate directions. Then $v(x) = v'(x - z)$ for $x \in \mathbb{R}^n$ is p-harmonic in $\mathbb{R}^n \setminus \Omega$. Comparing boundary values and using the maximum principle, as well as Lemmas 4.2, 4.3, we deduce that

$$c(1 - U(x)) \geq v(x) \quad \text{whenever } x \in B(0, 2\rho), \tag{13.57}$$

where c depends only on the data and the p-capacity of E. As in Proposition 13.5 we see that

$$c' \int_Q |\nabla U_+|^p \, d\mathcal{H}^{n-1} \geq (1 - U(y_Q))^p s(Q)^{n-1-p}. \tag{13.58}$$

Now from Krol's construction, we also deduce that

$$c'' v(y_Q) \geq s(Q)^{1-1/p}. \tag{13.59}$$

Combining (13.57)-(13.59) we conclude that for some $\tilde{c}$ with the same dependence as the above constants,

$$\tilde{c} \int_Q |\nabla U_+|^p \, d\mathcal{H}^{n-1} \geq s(Q)^{n-2}. \tag{13.60}$$

Now since $B(0, 4a) \cap P \subset E$ we see that for l large there are at least $\approx 2^{l(n-2)}$ members of $\{Q_k\}$ whose side length lies between $2^{-l-1}a$ and $2^{-l}a$. Using this fact and summing (13.60) we get Proposition 13.6 in this special case.

To get Proposition 13.6 for a general p-homogeneous f satisfying our structure conditions, we use a clever idea of Venouziou and Verchota in [**VV**]. To simplify matters we make a further translation, scaling and rotation, if necessary, so that E becomes E' with

$$\begin{aligned} &(a) \quad E' \subset B(0, \rho') \cap P \text{ for some } \rho' > 1, \\ &(b) \quad 0 \in \partial' E', \\ &(c) \quad E' \cap \partial'(B(0,1) \cap P) \neq \emptyset, \\ &(d) \quad E' \subset \{x \in \mathbb{R}^{n-1} : \ x_1 \leq 0\}. \end{aligned} \tag{13.61}$$

Then f, U, become f', U' under this transformation. Since f' satisfies the same structure assumptions as f it clearly suffices to prove Proposition 13.6 for E', U', the $\mathcal{A}' = \nabla f'$ capacitary function for E'. For ease of notation we drop the primes in (13.61) and just write U, E, f.

Let $D = B(0,1) \setminus E$ and given $t > 0$ for $0 < t \leq 1/8$, we set $z_t = \frac{1}{4} e_1 + t e_n$. Let

$$H_t = \{y + s(y - z_t) : s \geq 0 \text{ and } y \in E\} \cap \bar{B}(0,1) \quad \text{and} \quad D_t = B(0,1) \setminus H_t.$$

We note that H_t is the union of line segments with one endpoint on E and the other endpoint on $\partial B(0,1)$. Also, each line segment lies on a ray beginning at $(1/4, 0, \ldots, t)$ and D_t is a starlike Lipschitz domain with respect to z_t. That is,

∂D_t is the union of graphs of a finite number of Lipschitz functions defined on $(n-1)$-dimensional planes and if $y \in D_t$, then the ray joining z_t to y is also in D_t.

Let $\tilde{f}(\eta) = f(-\eta)$ whenever $\eta \in \mathbb{R}^n$ and let

$$\mathcal{G}_0 = \mathcal{G}_0(\cdot, z_t), \quad \mathcal{G}_1 = \mathcal{G}_1(\cdot, z_t), \quad \text{and} \quad \mathcal{G}_2 = \mathcal{G}_2(\cdot, z_t)$$

denote the $\tilde{\mathcal{A}} = \nabla \tilde{f}$-harmonic Green's functions for $B(0,1), D, D_t$ respectively with pole at z_t. Also, let $F = F(\cdot, z_t)$ be the $\tilde{\mathcal{A}}$-harmonic fundamental solution on $\mathbb{R}^n$ with pole at z_t. Put

$$\begin{cases} \zeta_0(\cdot, z_t) = F(\cdot, z_t) - \mathcal{G}_0(\cdot, z_t), \\ \zeta_1(\cdot, z_t) = F(\cdot, z_t) - \mathcal{G}_1(\cdot, z_t), \\ \zeta_2(\cdot, z_t) = F(\cdot, z_t) - \mathcal{G}_2(\cdot, z_t). \end{cases}$$

From (e) of Lemma 10.4 with $\mathcal{A}$ replaced by $\tilde{\mathcal{A}}$, we see that $\zeta_i = \zeta_i(\cdot, z_t)$ has a Hölder continuous extension to a neighborhood of z_t and so is locally Hölder continuous in its respective domain whenever $i \in \{0, 1, 2\}$.

To complete the proof of Proposition 13.6, we shall need the following Rellich-type identity.

LEMMA 13.7. *With the above notation, for* $i = 0, 2$,

$$\int_{\partial O} \langle x - z_t, \nu \rangle\, \tilde{f}(\nabla \mathcal{G}_i) d\mathcal{H}^{n-1} = \frac{(n-p)}{p(p-1)}\, \zeta_i(z_t) \tag{13.62}$$

where $O = B(0,1)$ *when* $i = 0$ *and* $O = D_t$ *when* $i = 2$ *with* ν *is the outer unit normal to* O.

PROOF. To start the proof of Lemma 13.7 we write $\mathcal{G}$ for $\mathcal{G}_i$ and use the Gauss-Green Theorem as in (11.10)-(11.14) with $B = \bar{B}(z_t, \epsilon), 0 < \epsilon$ small, to get

$$\begin{aligned} I &= \int_{O \setminus B} \nabla \cdot ((x - z_t) \tilde{f}(\nabla \mathcal{G}))\, dx \\ &= \int_{\partial O} \langle x - z_t, \nu \rangle \tilde{f}(\nabla \mathcal{G}) d\mathcal{H}^{n-1} + \int_{\partial B} \langle x - z_t, \nu \rangle \tilde{f}(\nabla \mathcal{G}) d\mathcal{H}^{n-1}. \end{aligned} \tag{13.63}$$

Moreover,

$$\begin{aligned} I &= n \int_{O \setminus B} \tilde{f}(\nabla \mathcal{G}) dx + \sum_{k,j=1}^{n} \int_{O \setminus B} (x_k - (z_t)_k) \tilde{f}_{\eta_j}(\nabla \mathcal{G}) \mathcal{G}_{x_k x_j} dx \\ &= n \int_{O \setminus B} \tilde{f}(\nabla \mathcal{G}) dx + I_1. \end{aligned} \tag{13.64}$$

Integrating I_1 by parts, using p-homogeneity of $\tilde{f}$, as well as $\tilde{\mathcal{A}} = \nabla \tilde{f}$-harmonicity of $\mathcal{G}$ in $O \setminus \bar{B}$, we deduce that

(13.65)

$$\begin{aligned} I_1 = &\int_{\partial O} \langle x - z_t, \nabla \mathcal{G} \rangle\, \langle \nabla \tilde{f}(\nabla \mathcal{G}), \nu \rangle\, d\mathcal{H}^{n-1} + \int_{\partial B} \langle x - z_t, \nabla \mathcal{G} \rangle\, \langle \nabla \tilde{f}(\nabla \mathcal{G}), \nu \rangle\, d\mathcal{H}^{n-1} \\ &- p \int_{O \setminus B} \tilde{f}(\nabla \mathcal{G}) dx. \end{aligned}$$

Combining (13.63)-(13.65), we find after some juggling that

$$
\begin{aligned}
(n-p)\int_{O\setminus B}\tilde f(\nabla\mathcal{G})dx = &\int_{\partial O}\langle x-z_t,\nu\rangle\tilde f(\nabla\mathcal{G})d\mathcal{H}^{n-1}\\
&-\int_{\partial O}\langle x-z_t,\nabla\mathcal{G}\rangle\,\langle\nabla\tilde f(\nabla\mathcal{G}),\nu\rangle\,d\mathcal{H}^{n-1}\\
&+\int_{\partial B}\langle x-z_t,\nu\rangle\tilde f(\nabla\mathcal{G})d\mathcal{H}^{n-1}\\
&-\int_{\partial B}\langle x-z_t,\nabla\mathcal{G}\rangle\,\langle\nabla\tilde f(\nabla\mathcal{G}),\nu\rangle\,d\mathcal{H}^{n-1}.
\end{aligned}
\tag{13.66}
$$

We note that $\nu=-\frac{\nabla\mathcal{G}}{|\nabla\mathcal{G}|}$ on ∂O for $\mathcal{H}^{n-1}$ almost everywhere. Using this fact and p-homogeneity of $\tilde f$, we obtain

$$
\begin{aligned}
\int_{\partial O}\langle x-z_t,\nu\rangle\tilde f(\nabla\mathcal{G})d\mathcal{H}^{n-1}-\int_{\partial O}\langle x-z_t,\nabla\mathcal{G}\rangle\,\langle\nabla\tilde f(\nabla\mathcal{G}),\nu\rangle\,d\mathcal{H}^{n-1}&\\
=-(p-1)\int_{\partial O}\langle x-z_t,\nu\rangle\tilde f(\nabla\mathcal{G})d\mathcal{H}^{n-1}.&
\end{aligned}
\tag{13.67}
$$

Now on ∂B we have $\nu=-\frac{x-z_t}{|x-z_t|}$ so

$$
\begin{aligned}
&\int_{\partial B}\langle x-z_t,\nu\rangle\tilde f(\nabla\mathcal{G})d\mathcal{H}^{n-1}-\int_{\partial B}\langle x-z_t,\nabla\mathcal{G}\rangle\,\langle\nabla\tilde f(\nabla\mathcal{G}),\nu\rangle\,d\mathcal{H}^{n-1}\\
&\quad=-\int_{\partial B}|x-z_t|\tilde f(\nabla\mathcal{G})dx+\int_{\partial B}\langle x-z_t,\nabla\mathcal{G}\rangle\,\langle\nabla\tilde f(\nabla\mathcal{G}),\tfrac{x-z_t}{|x-z_t|}\rangle d\mathcal{H}^{n-1}.
\end{aligned}
\tag{13.68}
$$

Again, from p-homogeneity of $\tilde f$, and $\tilde{\mathcal{A}}=\nabla\tilde f$-harmonicity of $\mathcal{G}$, and the Gauss-Green Theorem we see that

$$
p\int_{O\setminus B}\tilde f(\nabla\mathcal{G})dx=\int_{O\setminus B}\nabla\cdot(\mathcal{G}\nabla\tilde f(\nabla\mathcal{G}))dx=-\int_{\partial B}\mathcal{G}\,\langle\nabla\tilde f(\nabla\mathcal{G}),\tfrac{x-z_t}{|x-z_t|}\rangle d\mathcal{H}^{n-1}.
\tag{13.69}
$$

Using (13.67)-(13.69) in (13.66) we arrive after some more juggling at

$$
\begin{aligned}
&(n/p-1)\int_{\partial B}\mathcal{G}\langle\nabla\tilde f(\nabla\mathcal{G}),\tfrac{x-z_t}{|x-z_t|}\rangle\,d\mathcal{H}^{n-1}\\
&\quad+\int_{\partial B}\langle x-z_t,\nabla\mathcal{G}\rangle\,\langle\nabla\tilde f(\nabla\mathcal{G}),\tfrac{x-z_t}{|x-z_t|}\rangle\,d\mathcal{H}^{n-1}\\
&\quad-\int_{\partial B}|x-z_t|\tilde f(\nabla\mathcal{G})\,d\mathcal{H}^{n-1}=(p-1)\int_{\partial O}\langle x-z_t,\nu\rangle\tilde f(\nabla\mathcal{G})d\mathcal{H}^{n-1}.
\end{aligned}
\tag{13.70}
$$

We intend to let $\epsilon\to 0$ in the left-hand side of (13.70). To study the asymptotics as $\epsilon\to 0$, we note from Lemma 10.4 with $w=z_t$ and $\tilde{\mathcal{A}}=\mathcal{A}$ that $F-\mathcal{G}=\zeta$ in O where ζ is Hölder continuous in $B(z_t,1/8)$ for some exponent $\alpha\in(0,1)$ depending only on the data. Moreover, using (10.36), (10.42)-(10.43), elliptic regularity theory, and arguing as in (12.3)-(12.5) we see for some constant $c\geq 1$, depending only on the data that

$$
|\nabla\zeta(x)|\leq c\,\epsilon^{\alpha-1}\max_{\partial B(z_t,1/8)}\zeta\quad\text{whenever } x\in\partial B(z_t,\epsilon)\text{ and } 0<\epsilon\leq 1/100.
\tag{13.71}
$$

Also from the structure and regularity assumptions on $\tilde{f}$, we see that if $\eta', \eta^* \in B(\eta', |\eta'|/2)$ with $\eta' \neq 0$ then

$$(13.72)\qquad \begin{aligned} &(\alpha) \quad |\tilde{f}(\eta') - \tilde{f}(\eta^*)| \leq c|\eta'|^{p-1}|\eta' - \eta^*|, \\ &(\beta) \quad |\nabla \tilde{f}(\eta') - \nabla \tilde{f}(\eta^*)| \leq c|\eta'|^{p-2}|\eta' - \eta^*|. \end{aligned}$$

Moreover, from (10.36) we have

$$(13.73)\qquad |z - z_t|^{-1}|F(z)| + |\nabla F(z)| \approx |z - z_t|^{(1-n)/(p-1)} \quad \text{for } z \in \mathbb{R}^n \setminus \{z_t\}$$

where proportionality constants and c in (13.73) depend only on the data.

If $x \in \partial B(z_t, \epsilon)$ then from (13.71)-(13.73) with $z = x$ and $\eta^* = \nabla \mathcal{G}(x)$, $\eta' = \nabla F(x)$, we obtain

(13.74)

$$(\alpha') \quad |\tilde{f}(\nabla F(x)) - \tilde{f}(\nabla \mathcal{G}(x))| \leq \tilde{c}|\nabla F(x)|^{p-1} |\nabla \zeta(x)| \leq \tilde{c}^2|x - z_t|^{\alpha - n},$$

$$(\beta') \quad |\nabla \tilde{f}(\nabla F(x)) - \nabla \tilde{f}(\nabla \mathcal{G}(x))| \leq \tilde{c}|\nabla F(x)|^{p-2} |\nabla \zeta(x)| \leq \tilde{c}^2 \, |x - z_t|^{1-n+\alpha+\frac{(n-p)}{p-1}}.$$

Using (13.74), we find for the integrands on the left-hand side of (13.70) that

(13.75)

$$\begin{aligned} (n/p-1)\mathcal{G}\langle \nabla \tilde{f}(\nabla \mathcal{G}), \tfrac{x-z_t}{|x-z_t|}\rangle &= (n/p-1)(F-\zeta)\langle \nabla \tilde{f}(\nabla F), \tfrac{x!-z_t}{|x-z_t|}\rangle + O(|x-z_t|)^{1-n+\alpha}, \\ \langle x - z_t, \nabla \mathcal{G}\rangle \, \langle \nabla \tilde{f}(\nabla \mathcal{G}), \tfrac{x-z_t}{|x-z_t|}\rangle &= \langle x - z_t, \nabla F\rangle \, \langle \nabla \tilde{f}(\nabla F), \tfrac{x-z_t}{|x-z_t|}\rangle + O(|x - z_t|)^{1-n+\alpha}, \\ -|x - z_t|\tilde{f}(\nabla \mathcal{G}) &= -|x - z_t|\tilde{f}(\nabla F) + O(|x - z_t|^{1-n+\alpha}) \end{aligned}$$

where we have used standard big O notation. Replacing the sum of the left-hand side of these equations in (13.75) by the sum of the right-hand sides and using $(p-n)/(p-1)$-homogeneity of F in $x - z_t$, we get for $x \in \partial B(z_t, \epsilon)$ that

(13.76)

$$\begin{aligned} J_1 &= \int_{\partial B} \left\{ \tfrac{p-n}{p(p-1)} \, F(x) \, \langle \nabla \tilde{f}(\nabla F(x)), \tfrac{x-z_t}{|x-z_t|}\rangle - |x - z_t|\tilde{f}(\nabla F(x)) \right\} d\mathcal{H}^{n-1} \\ &= \int_{\partial B} \tfrac{n-p}{p} \, \zeta(x) \, \langle \nabla \tilde{f}(\nabla F(x)), \tfrac{x-z_t}{|x-z_t|} \rangle d\mathcal{H}^{n-1} \\ &\quad + (p-1)\int_{\partial O} \langle x - z_t, \nu \rangle \tilde{f}(\nabla \mathcal{G}(x)) d\mathcal{H}^{n-1} + O(\epsilon^\alpha) \\ &= J_2 + (p-1)\int_{\partial O} \langle x - z_t, \nu \rangle \tilde{f}(\nabla \mathcal{G}(x)) d\mathcal{H}^{n-1} + O(\epsilon^\alpha) \end{aligned}$$

where the last integral (integral over ∂O) is $p-1$ times the one we want to compute. We claim that

$$(13.77)\qquad J_1 \equiv 0.$$

To prove this claim we note from chapter 7 with f replaced by $\tilde{f}$ (see the sentence above (7.4)) that if $\tilde{f}(\eta) = k(-\eta)^p$ and

$$h(x) = \max\left\{ \frac{\langle x, y\rangle}{k(y)}, \; y \in \mathbb{R}^n \setminus \{0\} \right\},$$

then h has continuous second partials and

$$(13.78)\qquad \nabla k(\nabla h(x)) = x/h(x) \quad \text{while} \quad k(\nabla h) = 1.$$

Also it followed (see Remark 7.1) that $\tilde{F}(x) = \theta\,(h(x))^{(p-n)/(p-1)}$ is the fundamental solution for $\tilde{\mathcal{A}} = \nabla\tilde{f}$-harmonic functions with pole at zero where

$$\theta^{p-1} = p^{-1}\left(\frac{n-p}{p-1}\right)^{1-p}\left(\int_{\mathbb{S}^{n-1}} h(\omega)^{-n}d\omega\right)^{-1}.$$

Let $F(x) = \tilde{F}(x - z_t)$, whenever $x \in \mathbb{R}^n \setminus \{z_t\}$. Using these facts and (13.78) in J_1 and the homogeneity of the various functions, we see that

$$-|x-z_t|\,\tilde{f}(\nabla F(x)) = -\theta^p\,|x-z_t|\left(\frac{n-p}{p-1}\right)^p h(x-z_t)^{\frac{(1-n)p}{p-1}}. \tag{13.79}$$

Moreover, using (13.78), it follows that

$$\begin{aligned} F(x)\langle\nabla\tilde{f}(\nabla F(x)), \tfrac{x-z_t}{|x-z_t|}\rangle &= -\theta^p h(x-z_t)^{\frac{p-n}{p-1}} p\left(\frac{n-p}{p-1}\right)^{p-1} h(x-z_t)^{1-n} \\ &\quad\cdot\langle\nabla k(\nabla h(x-z_t)), \tfrac{x-z_t}{|x-z_t|}\rangle \\ &= -\theta^p p\left(\frac{n-p}{p-1}\right)^{p-1} h(x-z_t)^{\frac{(1-n)p}{p-1}}|x-z_t|. \end{aligned} \tag{13.80}$$

Multiplying the right-hand side of (13.80) by $\frac{p-n}{p(p-1)}$ and adding to (13.79) we obtain claim (13.77).

Next arguing as in (13.79), we get

$$J_2 = -\int_{\partial B(z_t,\epsilon)} \theta^{p-1}(n-p)\left(\frac{n-p}{p-1}\right)^{p-1}\zeta(x)|x-z_t|h(x-z_t)^{-n}d\mathcal{H}^{n-1}. \tag{13.81}$$

Using one homogeneity of h we see that (13.81) can be rewritten in spherical coordinates, $\epsilon\,\omega = x - z_t$, $\omega \in \mathbb{S}^{n-1}$, as

$$J_2 = -\int_{\mathbb{S}^{n-1}} \theta^{p-1}(n-p)\left(\frac{n-p}{p-1}\right)^{p-1}\zeta(z_t+\epsilon\omega)h(\omega)^{-n}d\mathcal{H}^{n-1}. \tag{13.82}$$

Letting $\epsilon \to 0$ and using continuity of ζ at z_t we conclude from (13.76), (13.77), and (13.82) that (13.62) in Lemma 13.7 is true since

$$\theta^{p-1}\left(\frac{n-p}{p-1}\right)^p\int_{\mathbb{S}^{n-1}} h(\omega)^{-n}d\mathcal{H}^{n-1} = \frac{n-p}{p(p-1)}. \tag{13.83}$$

This finishes the proof of Lemma 13.7. □

Proof of Proposition 13.6. We shall apply Lemma 13.7 with $O = D_t$. Before doing this we note that if ν is the outer unit normal to D_t then $\langle x - z_t, \nu(x)\rangle = 0$ when $x \in \partial D_t \setminus \mathbb{S}^{n-1}$ and $x_n < 0$ while $\langle x - z_t, \nu(x)\rangle = t$ when $x \in E \cap B(0,1)$ for $\mathcal{H}^{n-1}$-almost everywhere. Using these facts and Lemma 13.7 we obtain for

$$\mathcal{G}_2 = \mathcal{G}_2(\cdot, z_t) \quad\text{and}\quad \zeta_2 = \zeta_2(\cdot, z_t)$$

that

$$\begin{aligned} \gamma\,\zeta_2(z_t) &= \int_{\partial D_t}\langle x-z_t,\nu\rangle\tilde{f}(\nabla\mathcal{G}_2)d\mathcal{H}^{n-1} \\ &= t\int_{E\cap B(0,1)}\tilde{f}(\nabla\mathcal{G}_2)d\mathcal{H}^{n-1} + \int_{\partial D_t\cap\mathbb{S}^{n-1}}\langle x-z_t,\nu\rangle\tilde{f}(\nabla\mathcal{G}_2)d\mathcal{H}^{n-1} \end{aligned} \tag{13.84}$$

where $\gamma := \frac{n-p}{p(p-1)}$ is the constant in (13.83). Now from the maximum principle for $\tilde{\mathcal{A}}$-harmonic functions we have for

$$\mathcal{G}_i = \mathcal{G}_i(\cdot, z_t), \quad \text{and} \quad \zeta_i = \zeta_i(\cdot, z_t) \quad \text{for } i = 0, 1, 2,$$

that

$$\mathcal{G}_2 \leq \mathcal{G}_1 \leq \mathcal{G}_0 \text{ in } D_t \quad \text{so} \quad \zeta_0(z_t) \leq \zeta_1(z_t) \leq \zeta_2(z_t). \tag{13.85}$$

Using (13.85), the mean value theorem, the fact that $\nabla \mathcal{G}_i, i = 1, 2$, has non-tangential limits from above $\mathcal{H}^{n-1}$-almost everywhere on E and that all limits have the same direction, we conclude that

$$\tilde{f}(\nabla \mathcal{G}_2) \leq \tilde{f}((\nabla \mathcal{G}_1)_+) \quad \text{on } E \cap B(0,1).$$

Likewise,

$$\tilde{f}(\nabla \mathcal{G}_2) \leq \tilde{f}(\nabla \mathcal{G}_0) \quad \text{for } \mathcal{H}^{n-1} \text{ almost every } x \in \mathbb{S}^{n-1} \cap \partial D_t.$$

Using these facts and (13.84)-(13.85), and Lemma 13.7 with $O = B(0,1)$, we get

$$\begin{aligned} t\int_{E\cap B(0,1)} \tilde{f}((\nabla \mathcal{G}_1)_+) d\mathcal{H}^{n-1} &\geq \gamma\zeta_2(z_t) - \int_{\mathbb{S}^{n-1}} \langle x - z_t, \nu\rangle \tilde{f}(\nabla \mathcal{G}_0) d\mathcal{H}^{n-1} \\ &= \gamma(\zeta_2(z_t) - \zeta_0(z_t)) \\ &\geq \gamma(\zeta_1(z_t) - \zeta_0(z_t)). \end{aligned} \tag{13.86}$$

Letting $t \to 0$ in (13.86) we assert that to complete the proof of Proposition 13.6 it suffices to show

$$c\,[\zeta_1(z_t) - \zeta_0(z_t)] \geq 1 \tag{13.87}$$

where $c \geq 1$ is a positive constant depending on the data and $\text{Cap}_{\mathcal{A}}(E)$ but independent of t. Indeed, from (4.4) (b) of Lemma 4.2 and (10.31) (c) of Lemma 10.4, as well as the maximum principle for $\tilde{\mathcal{A}}$-harmonic functions, we find $c \geq 1$, independent of t, with $c\,(1 - U) \geq \mathcal{G}_1$ on $D_t \setminus B(e_1/4, 1/8)$ when $0 < t < 1/100$. Then as in the displays above (13.86) if follows that

$$c\tilde{f}(-(\nabla U_+)) = cf(\nabla U_+) \geq \tilde{f}((\nabla \mathcal{G}_1)_+)$$

for $\mathcal{H}^{n-1}$ almost every $x \in E \cap B(0,1)$. Using this inequality, (13.87), and letting $t \to 0$ in (13.86) we get Proposition 13.6.

To prove (13.87) we note as in Lemma 3.3 that for some $\tilde{\alpha} \in (0,1), c' \geq 1$, we have

$$\max_{B(0,s)} \mathcal{G}_1 \leq c' s^{\tilde{\alpha}} \mathcal{G}_1(e_1/16) \quad \text{for } 0 < s \leq 1/16$$

where c' depends only on the data and $\text{Cap}_{\mathcal{A}}(E)$. Also from Lemma 3.1 we see that $c'' \min_{\bar{B}(0,1/16)} \mathcal{G}_0 \geq 1$ where c'' depends only on the data. Thus there exists $\hat{\rho}, 0 < \hat{\rho} < 1/16$ with the same dependence as c' such that

$$\mathcal{G}_0 - \mathcal{G}_1 \geq \hat{\rho} \quad \text{in } \bar{B}(0, \hat{\rho}) \setminus E. \tag{13.88}$$

We claim that

$$|\nabla \mathcal{G}_0(x)| \approx \frac{\mathcal{G}_0(x)}{|x - z_t|} \approx |x - z_t|^{\frac{1-n}{p-1}} \quad \text{for } x \in B(0, 1/2). \tag{13.89}$$

The left-hand inequality in (13.89) follows from (10.44) (β) while the right-hand inequality is a consequence of Lemma 10.4 (e) and our knowledge of F.

Armed with (13.88), (13.89), we now prove (13.87), and so also Proposition 13.6. From (13.89) it follows in a now well-known way that $\mathcal{G}_0 - \mathcal{G}_1$ satisfies a locally uniformly elliptic PDE in $B(0,1/2) \setminus E$ similar to (10.37), (10.38). From Harnack's inequality for this PDE and (13.88) we deduce for some $\tilde{\rho} > 0$ that

$$\mathcal{G}_0 - \mathcal{G}_1 \geq \tilde{\rho} \quad \text{on } \partial B(e_1/4, 1/4 - \hat{\rho}).$$

Using the same argument as in the proof of Lemma 10.4 (e), it now follows that $\zeta_1(z_t) - \zeta_0(z_t) \geq \tilde{\rho}$ whenever $t \in (0, 1/8)$. We conclude that (13.87) and Proposition 13.6 are true.

We next show that if E has empty interior, the assumption (13.44) holds. To this end, we consider following three cases depending on p. For $1 < p < n-1$, we note from (13.38) and (2.6) that for some $C_+ \geq 1$ independent of j,

$$C_+^{-1} \operatorname{diam}(E_j) \leq \operatorname{Cap}_{\mathcal{A}}(E_j) \leq C_+ \left(\operatorname{diam}(E_j)\right)^{n-p}. \tag{13.90}$$

Thus, (13.90) implies that $\operatorname{diam}(E_j)$ is bounded below independently of j, which in view of (13.41) implies that (13.44) always holds when $1 < p < n-1$. When $p = n-1$, then from (13.41) we deduce that $\operatorname{Cap}_{\mathcal{A}}(E) = 1$ so again (13.44) always holds. Finally, when $n - 1 < p < n$, then a line segment of length l has capacity $\approx l^{n-p}$ so from (13.41) we deduce that if (13.44) is false then $\operatorname{diam}(E_j) \to 0$. For $j = 1, 2, \ldots$, choose $s_j, \hat{E}_j$, so that $s_j \hat{E}_j = E_j$ and $\operatorname{Cap}_{\mathcal{A}}(\hat{E}_j) = 1$. Then from the above discussion we find that $s_j \to 0$ as $j \to \infty$. Let $\hat{\mu}_j$ denote the measure in Theorem B defined relative to $\hat{E}_j$. Then from the usual dilation argument we have $\hat{\mu}_j = s_j^{p+1-n} \mu_j$. From (13.34) we see that $\{\hat{\mu}_j\}_{j \geq 1}$ converges weakly to zero and a subsequence of $\{\hat{E}_j\}$ converges to $\hat{E}$ a set of $\mathcal{A}$-capacity one. We can now argue as previously with $\hat{E}$ replacing E to get a contradiction. Using just uniform boundedness of $\{\hat{\mu}_j\}_{j \geq 1}$ and our earlier work it follows that E has nonempty interior. From weak convergence of measures in Proposition 11.1 we now get a contradiction since $\hat{\mu}_j \to 0$ weakly as $j \to \infty$. Thus, assumption (13.44) holds when $1 < p < n$. The proof of existence in Theorem B is now complete. □

13.3. Uniqueness of Minkowski problem

Uniqueness in Theorem B can be shown using the equality result in the Brunn-Minkowski inequality(Theorem A) as in [**CNSXYZ**] or [**CJL**]:

PROOF OF $(c), (e)$ IN THEOREM B. To prove uniqueness in Theorem B, suppose μ is a positive finite Borel measure on $\mathbb{S}^{n-1}$ satisfying (8.1) and let E_0, E_1 be two compact convex sets with nonempty interiors satisfying (8.5) in Theorem B relative to μ. Let h_0 and h_1 be the support functions of E_0 and E_1 respectively. For $t \in [0,1]$ we let $E_t = (1-t)E_0 + tE_1$. Using Proposition 12.1 and (12.31) we deduce as in (13.30) and (13.31) that if $p \neq n-1$, then

$$\begin{aligned} \frac{d}{dt} \operatorname{Cap}_{\mathcal{A}}(E_t)\bigg|_{t=0} &= (p-1) \int_{\mathbb{S}^{n-1}} (h_1(\xi) - h_0(\xi)) d\mu(\xi) \\ &= (n-p)[\operatorname{Cap}_{\mathcal{A}}(E_1) - \operatorname{Cap}_{\mathcal{A}}(E_0)]. \end{aligned} \tag{13.91}$$

We define

$$\mathbf{m}(t) = \operatorname{Cap}_{\mathcal{A}}(E_t)^{\frac{1}{n-p}}.$$

Then basic calculus and (13.91) gives us that

$$\begin{aligned} \mathbf{m}'(0) &= \mathrm{Cap}_{\mathcal{A}}(E_0)^{\frac{1}{n-p}-1}[\mathrm{Cap}_{\mathcal{A}}(E_1) - \mathrm{Cap}_{\mathcal{A}}(E_0)] \\ &= \mathbf{m}(0)^{1-n+p}[\mathbf{m}(1)^{n-p} - \mathbf{m}(0)^{n-p}]. \end{aligned} \tag{13.92}$$

From (2.7) in Theorem A with E_1, E_2, λ replaced by E_0, E_1, t we find that $\mathbf{m}$ is a concave function on $[0,1]$ and therefore

$$\mathbf{m}'(0) \geq \mathbf{m}(1) - \mathbf{m}(0) \tag{13.93}$$

with strict inequality unless $\mathbf{m}$ is linear in t, which implies equality holds in the Brunn Minkowski inequality for $t \in [0,1]$. Let

$$l = \left(\frac{\mathrm{Cap}_{\mathcal{A}}(E_1)}{\mathrm{Cap}_{\mathcal{A}}(E_0)} \right)^{\frac{1}{n-p}}.$$

Using (13.93) in (13.92) we see that

$$l^{n-p} - 1 \geq l - 1.$$

Reversing the roles of E_0, E_1 we also get

$$l^{p-n} - 1 \geq l^{-1} - 1.$$

Clearly, both these inequalities can only hold if $l = 1$. Thus $\mathrm{Cap}_{\mathcal{A}}(E_0) = \mathrm{Cap}_{\mathcal{A}}(E_1)$ and equality holds in (2.7) for $t \in [0,1]$. From Theorem A we conclude that E_0 is a translate and dilate of E_1. From (2.5) it follows that honest dilations are not possible when $p \neq n-1$.

If $p = n-1$, let b_0, b_1 correspond to E_0, E_1, respectively as in (8.5) (d). Then $\mathrm{Cap}_{\mathcal{A}}(E_i) = 1$, for $i = 0, 1$ and arguing as in (13.91) we see that

(13.94)

$$b_0 \frac{d}{dt} \mathrm{Cap}_{\mathcal{A}}(E_t)\bigg|_{t=0} = (p-1) \int_{\mathbb{S}^{n-1}} (h_1(\xi) - h_0(\xi)) d\mu(\xi) = b_1 \frac{d}{dt} \mathrm{Cap}_{\mathcal{A}}(E_t)\bigg|_{t=1}.$$

From concavity of $\mathbf{m}(t)$ as above we see that $\mathbf{m}'(0) \geq \mathbf{m}'(1)$ so (13.94) implies $b_0 \leq b_1$ with strict inequality unless equality holds in (2.7) of Theorem A for E_t, $t \in [0,1]$. Reversing the roles of E_0, E_1 we get that $b_0 = b_1$ so from Theorem A, E_1 is homothetic to E_0.

This finishes the proof of (c), (e). in Theorem B and so also of Theorem B. □

Acknowledgment

This material is based upon work supported by National Science Foundation under Grant No. DMS-1440140 while the first author were in residence at the MSRI in Berkeley, California, during the Spring 2017 semester. The first author was also supported by ICMAT Severo Ochoa project SEV-2015-0554, and also acknowledges that the research leading to these results has received funding from the European Research Council under the European Union's Seventh Framework Programme (FP7/2007-2013)/ERC agreement no. 615112 HAPDEGMT. The fourth author was partially supported by NSF DMS-1265996.

Bibliography

[A] M. Akman, *On the dimension of a certain measure in the plane*, Ann. Acad. Sci. Fenn. Math. **39** (2014), no. 1, 187–209, DOI 10.5186/aasfm.2014.3923. MR3186813

[A1] A. D. Aleksandrov. On the theory of mixed volumes. iii. extension of two theorems of minkowski on convex polyhedra to arbitrary convex bodie*s. *Mat. Sb. (N.S.)*, 3:27–46, 1938.

[A2] A. D. Aleksandrov. On the surface area measure of convex bodies. *Mat. Sb. (N.S.)*, 6:167–174, 1939.

[ALV] M. Akman, J. Lewis, and A. Vogel, *σ-finiteness of elliptic measures for quasilinear elliptic PDE in space*, Adv. Math. **309** (2017), 512–557, DOI 10.1016/j.aim.2017.01.013. MR3607285

[B1] C. Borell, *Capacitary inequalities of the Brunn-Minkowski type*, Math. Ann. **263** (1983), no. 2, 179–184, DOI 10.1007/BF01456879. MR698001

[B2] C. Borell, *Hitting probabilities of killed Brownian motion: a study on geometric regularity*, Ann. Sci. École Norm. Sup. (4) **17** (1984), no. 3, 451–467. MR777379

[BGMX] B. Bian, P. Guan, X.-N. Ma, and L. Xu, *A constant rank theorem for quasiconcave solutions of fully nonlinear partial differential equations*, Indiana Univ. Math. J. **60** (2011), no. 1, 101–119, DOI 10.1512/iumj.2011.60.4222. MR2952411

[BLS] C. Bianchini, M. Longinetti, and P. Salani, *Quasiconcave solutions to elliptic problems in convex rings*, Indiana Univ. Math. J. **58** (2009), no. 4, 1565–1589, DOI 10.1512/iumj.2009.58.3539. MR2542973

[C] A. Colesanti, *Brunn-Minkowski inequalities for variational functionals and related problems*, Adv. Math. **194** (2005), no. 1, 105–140, DOI 10.1016/j.aim.2004.06.002. MR2141856

[CC] A. Colesanti and P. Cuoghi, *The Brunn-Minkowski inequality for the n-dimensional logarithmic capacity of convex bodies*, Potential Anal. **22** (2005), no. 3, 289–304, DOI 10.1007/s11118-004-1326-7. MR2134723

[CF] R. R. Coifman and C. Fefferman, *Weighted norm inequalities for maximal functions and singular integrals*, Studia Math. **51** (1974), 241–250, DOI 10.4064/sm-51-3-241-250. MR0358205

[CFMS] L. Caffarelli, E. Fabes, S. Mortola, and S. Salsa, *Boundary behavior of nonnegative solutions of elliptic operators in divergence form*, Indiana Univ. Math. J. **30** (1981), no. 4, 621–640, DOI 10.1512/iumj.1981.30.30049. MR620271

[CJL] L. A. Caffarelli, D. Jerison, and E. H. Lieb, *On the case of equality in the Brunn-Minkowski inequality for capacity*, Adv. Math. **117** (1996), no. 2, 193–207, DOI 10.1006/aima.1996.0008. MR1371649

[CNSXYZ] A. Colesanti, K. Nyström, P. Salani, J. Xiao, D. Yang, and G. Zhang, *The Hadamard variational formula and the Minkowski problem for p-capacity*, Adv. Math. **285** (2015), 1511–1588, DOI 10.1016/j.aim.2015.06.022. MR3406534

[CS] A. Colesanti and P. Salani, *The Brunn-Minkowski inequality for p-capacity of convex bodies*, Math. Ann. **327** (2003), no. 3, 459–479, DOI 10.1007/s00208-003-0460-7. MR2021025

[CS1] A. Cianchi and P. Salani, *Overdetermined anisotropic elliptic problems*, Math. Ann. **345** (2009), no. 4, 859–881, DOI 10.1007/s00208-009-0386-9. MR2545870

[D] B. E. J. Dahlberg, *Estimates of harmonic measure*, Arch. Rational Mech. Anal. **65** (1977), no. 3, 275–288, DOI 10.1007/BF00280445. MR0466593

[EG] L. C. Evans and R. F. Gariepy, *Measure theory and fine properties of functions*, Studies in Advanced Mathematics, CRC Press, Boca Raton, FL, 1992. MR1158660

[FJ] W. Fenchel and B. Jessen. Mengenfunktionen und konvexe körper, danske vid. selsk. *Mat.-Fys. Medd.*, 16:1–31, 1938.

[G] R. J. Gardner, *The Brunn-Minkowski inequality*, Bull. Amer. Math. Soc. (N.S.) **39** (2002), no. 3, 355–405, DOI 10.1090/S0273-0979-02-00941-2. MR1898210

[GL] N. Garofalo and F.-H. Lin, *Unique continuation for elliptic operators: a geometric-variational approach*, Comm. Pure Appl. Math. **40** (1987), no. 3, 347–366, DOI 10.1002/cpa.3160400305. MR882069

[GT] D. Gilbarg and N. S. Trudinger, *Elliptic partial differential equations of second order*, Classics in Mathematics, Springer-Verlag, Berlin, 2001. Reprint of the 1998 edition. MR1814364

[Ga] R. M. Gabriel, *An extended principle of the maximum for harmonic functions in 3-dimensions*, J. London Math. Soc. **30** (1955), 388–401, DOI 10.1112/jlms/s1-30.4.388. MR0072959

[HKM] J. Heinonen, T. Kilpeläinen, and O. Martio, *Nonlinear potential theory of degenerate elliptic equations*, Dover Publications, Inc., Mineola, NY, 2006. Unabridged republication of the 1993 original. MR2305115

[J] D. Jerison, *A Minkowski problem for electrostatic capacity*, Acta Math. **176** (1996), no. 1, 1–47, DOI 10.1007/BF02547334. MR1395668

[K] N. J. Korevaar, *Convexity of level sets for solutions to elliptic ring problems*, Comm. Partial Differential Equations **15** (1990), no. 4, 541–556, DOI 10.1080/03605309908820698. MR1046708

[KP] C. E. Kenig and J. Pipher, *The Dirichlet problem for elliptic equations with drift terms*, Publ. Mat. **45** (2001), no. 1, 199–217, DOI 10.5565/PUBLMAT_45101_09. MR1829584

[KZ] T. Kilpeläinen and X. Zhong, *Growth of entire $\mathcal{A}$-subharmonic functions*, Ann. Acad. Sci. Fenn. Math. **28** (2003), no. 1, 181–192. MR1976839

[Kr] I. N. Krol′, *The behavior of the solutions of a certain quasilinear equation near zero cusps of the boundary* (Russian), Trudy Mat. Inst. Steklov. **125** (1973), 140–146, 233. Boundary value problems of mathematical physics, 8. MR0344671

[L] J. L. Lewis, *Capacitary functions in convex rings*, Arch. Rational Mech. Anal. **66** (1977), no. 3, 201–224, DOI 10.1007/BF00250671. MR0477094

[LLN] J. L. Lewis, N. Lundström, and K. Nyström, *Boundary Harnack inequalities for operators of p-Laplace type in Reifenberg flat domains*, Perspectives in partial differential equations, harmonic analysis and applications, Proc. Sympos. Pure Math., vol. 79, Amer. Math. Soc., Providence, RI, 2008, pp. 229–266, DOI 10.1090/pspum/079/2500495. MR2500495

[LN] J. L. Lewis and K. Nyström, *Boundary behaviour for p harmonic functions in Lipschitz and starlike Lipschitz ring domains* (English, with English and French summaries), Ann. Sci. École Norm. Sup. (4) **40** (2007), no. 5, 765–813, DOI 10.1016/j.ansens.2007.09.001. MR2382861

[LN1] J. Lewis and K. Nyström, *Boundary behavior and the Martin boundary problem for p harmonic functions in Lipschitz domains*, Ann. of Math. (2) **172** (2010), no. 3, 1907–1948, DOI 10.4007/annals.2010.172.1907. MR2726103

[LN2] J. L. Lewis and K. Nyström, *Regularity and free boundary regularity for the p Laplacian in Lipschitz and C^1 domains*, Ann. Acad. Sci. Fenn. Math. **33** (2008), no. 2, 523–548. MR2431379

[LN3] J. L. Lewis and K. Nyström, *Regularity and free boundary regularity for the p-Laplace operator in Reifenberg flat and Ahlfors regular domains*, J. Amer. Math. Soc. **25** (2012), no. 3, 827–862, DOI 10.1090/S0894-0347-2011-00726-1. MR2904575

[LN4] J. L. Lewis and K. Nyström, *Quasi-linear PDEs and low-dimensional sets*, J. Eur. Math. Soc. (JEMS) **20** (2018), no. 7, 1689–1746, DOI 10.4171/JEMS/797. MR3807311

[LSW] W. Littman, G. Stampacchia, and H. F. Weinberger, *Regular points for elliptic equations with discontinuous coefficients*, Ann. Scuola Norm. Sup. Pisa (3) **17** (1963), 43–77. MR0161019

[Li] G. M. Lieberman, *Boundary regularity for solutions of degenerate elliptic equations*, Nonlinear Anal. **12** (1988), no. 11, 1203–1219, DOI 10.1016/0362-546X(88)90053-3. MR969499

[Lo] M. Longinetti, *Some isoperimetric inequalities for the level curves of capacity and Green's functions on convex plane domains*, SIAM J. Math. Anal. **19** (1988), no. 2, 377–389, DOI 10.1137/0519028. MR930034

[M1] H. Minkowski, *Volumen und Oberfläche* (German), Math. Ann. **57** (1903), no. 4, 447–495, DOI 10.1007/BF01445180. MR1511220

[M2] Hermann Minkowski. Allgemeine lehrsätze über die convexen polyeder. *Nachrichten von der Gesellschaft der Wissenschaften zu Göttingen, Mathematisch-Physikalische Klasse*, 1897:198–220, 1897.

[S] J. Serrin, *Local behavior of solutions of quasi-linear equations*, Acta Math. **111** (1964), 247–302, DOI 10.1007/BF02391014. MR0170096

[Sc] R. Schneider, *Convex bodies: the Brunn-Minkowski theory*, Encyclopedia of Mathematics and its Applications, vol. 44, Cambridge University Press, Cambridge, 1993, DOI 10.1017/CBO9780511526282. MR1216521

[St] E. M. Stein, *Singular integrals and differentiability properties of functions*, Princeton Mathematical Series, No. 30, Princeton University Press, Princeton, N.J., 1970. MR0290095

[T] P. Tolksdorf, *Regularity for a more general class of quasilinear elliptic equations*, J. Differential Equations **51** (1984), no. 1, 126–150, DOI 10.1016/0022-0396(84)90105-0. MR727034

[VV] M. Venouziou and G. C. Verchota, *The mixed problem for harmonic functions in polyhedra of* $\mathbb{R}^3$, Perspectives in partial differential equations, harmonic analysis and applications, Proc. Sympos. Pure Math., vol. 79, Amer. Math. Soc., Providence, RI, 2008, pp. 407–423, DOI 10.1090/pspum/079/2500502. MR2500502

Editorial Information

To be published in the *Memoirs*, a paper must be correct, new, nontrivial, and significant. Further, it must be well written and of interest to a substantial number of mathematicians. Piecemeal results, such as an inconclusive step toward an unproved major theorem or a minor variation on a known result, are in general not acceptable for publication.

Papers appearing in *Memoirs* are generally at least 80 and not more than 200 published pages in length. Papers less than 80 or more than 200 published pages require the approval of the Managing Editor of the Transactions/Memoirs Editorial Board. Published pages are the same size as those generated in the style files provided for $\mathcal{A}\mathcal{M}\mathcal{S}$-LaTeX or $\mathcal{A}\mathcal{M}\mathcal{S}$-TeX.

Information on the backlog for this journal can be found on the AMS website starting from `http://www.ams.org/memo`.

A Consent to Publish is required before we can begin processing your paper. After a paper is accepted for publication, the Providence office will send a Consent to Publish and Copyright Agreement to all authors of the paper. By submitting a paper to the *Memoirs*, authors certify that the results have not been submitted to nor are they under consideration for publication by another journal, conference proceedings, or similar publication.

Information for Authors

Memoirs is an author-prepared publication. Once formatted for print and on-line publication, articles will be published as is with the addition of AMS-prepared frontmatter and backmatter. Articles are not copyedited; however, confirmation copy will be sent to the authors.

Initial submission. The AMS uses Centralized Manuscript Processing for initial submissions. Authors should submit a PDF file using the Initial Manuscript Submission form found at `www.ams.org/submission/memo`, or send one copy of the manuscript to the following address: Centralized Manuscript Processing, MEMOIRS OF THE AMS, 201 Charles Street, Providence, RI 02904-2294 USA. If a paper copy is being forwarded to the AMS, indicate that it is for *Memoirs* and include the name of the corresponding author, contact information such as email address or mailing address, and the name of an appropriate Editor to review the paper (see the list of Editors below).

The paper must contain a *descriptive title* and an *abstract* that summarizes the article in language suitable for workers in the general field (algebra, analysis, etc.). The *descriptive title* should be short, but informative; useless or vague phrases such as "some remarks about" or "concerning" should be avoided. The *abstract* should be at least one complete sentence, and at most 300 words. Included with the footnotes to the paper should be the 2020 *Mathematics Subject Classification* representing the primary and secondary subjects of the article. The classifications are accessible from `www.ams.org/msc/`. The Mathematics Subject Classification footnote may be followed by a list of *key words and phrases* describing the subject matter of the article and taken from it. Journal abbreviations used in bibliographies are listed in the latest *Mathematical Reviews* annual index. The series abbreviations are also accessible from `www.ams.org/msnhtml/serials.pdf`. To help in preparing and verifying references, the AMS offers MR Lookup, a Reference Tool for Linking, at `www.ams.org/mrlookup/`.

Electronically prepared manuscripts. The AMS encourages electronically prepared manuscripts, with a strong preference for $\mathcal{A}\mathcal{M}\mathcal{S}$-LaTeX. To this end, the Society has prepared $\mathcal{A}\mathcal{M}\mathcal{S}$-LaTeX author packages for each AMS publication. Author packages include instructions for preparing electronic manuscripts, samples, and a style file that generates the particular design specifications of that publication series. Though $\mathcal{A}\mathcal{M}\mathcal{S}$-LaTeX is the highly preferred format of TeX, author packages are also available in $\mathcal{A}\mathcal{M}\mathcal{S}$-TeX.

Authors may retrieve an author package for *Memoirs of the AMS* from `www.ams.org/journals/memo/memoauthorpac.html`. The *AMS Author Handbook* is available in PDF format from the author package link. The author package can also be obtained free of charge by sending email to tech-support@ams.org or from the Publication Division, American Mathematical Society, 201 Charles St., Providence, RI 02904-2294, USA. When requesting an author package, please specify $\mathcal{A}\mathcal{M}\mathcal{S}$-LaTeX or $\mathcal{A}\mathcal{M}\mathcal{S}$-TeX and the publication in which your paper will appear. Please be sure to include your complete mailing address.

After acceptance. The source files for the final version of the electronic manuscript should be sent to the Providence office immediately after the paper has been accepted for publication. The author should also submit a PDF of the final version of the paper to the editor, who will forward a copy to the Providence office.

Accepted electronically prepared files can be submitted via the web at www.ams.org/submit-book-journal/, sent via FTP, or sent on CD to the Electronic Prepress Department, American Mathematical Society, 201 Charles Street, Providence, RI 02904-2294 USA. TeX source files and graphic files can be transferred over the Internet by FTP to the Internet node ftp.ams.org (130.44.1.100). When sending a manuscript electronically via CD, please be sure to include a message indicating that the paper is for the *Memoirs*.

Electronic graphics. Comprehensive instructions on preparing graphics are available at www.ams.org/authors/journals.html. A few of the major requirements are given here.

Submit files for graphics as EPS (Encapsulated PostScript) files. This includes graphics originated via a graphics application as well as scanned photographs or other computer-generated images. If this is not possible, TIFF files are acceptable as long as they can be opened in Adobe Photoshop or Illustrator.

Authors using graphics packages for the creation of electronic art should also avoid the use of any lines thinner than 0.5 points in width. Many graphics packages allow the user to specify a “hairline” for a very thin line. Hairlines often look acceptable when proofed on a typical laser printer. However, when produced on a high-resolution laser imagesetter, hairlines become nearly invisible and will be lost entirely in the final printing process.

Screens should be set to values between 15% and 85%. Screens which fall outside of this range are too light or too dark to print correctly. Variations of screens within a graphic should be no less than 10%.

Any graphics created in color will be rendered in grayscale for the printed version unless color printing is authorized by the Managing Editor and the Publisher. In general, color graphics will appear in color in the online version.

Inquiries. Any inquiries concerning a paper that has been accepted for publication should be sent to memo-query@ams.org or directly to the Electronic Prepress Department, American Mathematical Society, 201 Charles St., Providence, RI 02904-2294 USA.

Editors

SELECTED PUBLISHED TITLES IN THIS SERIES

1323 **Shu Kawaguchi and Kazuhiko Yamaki,** Effective Faithful Tropicalizations Associated to Linear Systems on Curves, 2021

1322 **D. Bulacu and B. Torrecillas,** Galois and Cleft Monoidal Cowreaths. Applications, 2021

1321 **Christian Haase, Andreas Paffenholz, Lindsey C. Piechnik, and Francisco Santos,** Existence of Unimodular Triangulations–Positive Results, 2021

1320 **Th. Heidersdorf and R. Weissauer,** Cohomological Tensor Functors on Representations of the General Linear Supergroup, 2021

1319 **Abed Bounemoura and Jacques Féjoz,** Hamiltonian Perturbation Theory for Ultra-Differentiable Functions, 2021

1318 **Chao Wang, Zhifei Zhang, Weiren Zhao, and Yunrui Zheng,** Local Well-Posedness and Break-Down Criterion of the Incompressible Euler Equations with Free Boundary, 2021

1317 **Eric M. Rains and S. Ole Warnaar,** Bounded Littlewood Identities, 2021

1316 **Ulrich Bunke and David Gepner,** Differential Function Spectra, the Differential Becker-Gottlieb Transfer, and Applications to Differential Algebraic K-Theory, 2021

1315 **S. Grivaux, É. Matheron, and Q. Menet,** Linear Dynamical Systems on Hilbert Spaces: Typical Properties and Explicit Examples, 2021

1314 **Pierre Albin, Frédéric Rochon, and David Sher,** Resolvent, Heat Kernel, and Torsion under Degeneration to Fibered Cusps, 2021

1313 **Paul Godin,** The 2D Compressible Euler Equations in Bounded Impermeable Domains with Corners, 2021

1312 **Patrick Delorme, Pascale Harinck, and Yiannis Sakellaridis,** Paley-Wiener Theorems for a p-Adic Spherical Variety, 2021

1311 **Lyudmila Korobenko, Cristian Rios, Eric Sawyer, and Ruipeng Shen,** Local Boundedness, Maximum Principles, and Continuity of Solutions to Infinitely Degenerate Elliptic Equations with Rough Coefficients, 2021

1310 **Hiroshi Iritani, Todor Milanov, Yongbin Ruan, and Yefeng Shen,** Gromov-Witten Theory of Quotients of Fermat Calabi-Yau Varieties, 2021

1309 **Jérémie Chalopin, Victor Chepoi, Hiroshi Hirai, and Damian Osajda,** Weakly Modular Graphs and Nonpositive Curvature, 2020

1308 **Christopher L. Douglas, Christopher Schommer-Pries, and Noah Snyder,** Dualizable Tensor Categories, 2020

1307 **Adam R. Thomas,** The Irreducible Subgroups of Exceptional Algebraic Groups, 2020

1306 **Kazuyuki Hatada,** Hecke Operators and Systems of Eigenvalues on Siegel Cusp Forms, 2020

1305 **Bogdan Ion and Siddhartha Sahi,** Double Affine Hecke Algebras and Congruence Groups, 2020

1304 **Matthias Fischmann, Andreas Juhl, and Petr Somberg,** Conformal Symmetry Breaking Differential Operators on Differential Forms, 2020

1303 **Zhi Qi,** Theory of Fundamental Bessel Functions of High Rank, 2020

1302 **Paul M. N. Feehan and Manousos Maridakis,** Łojasiewicz-Simon Gradient Inequalities for Coupled Yang-Mills Energy Functionals, 2020

1301 **Joachim Krieger,** On Stability of Type II Blow Up for the Critical Nonlinear Wave Equation in $\mathbb{R}^{3+1}$, 2020

1300 **Camille Male,** Traffic Distributions and Independence: Permutation Invariant Random Matrices and the Three Notions of Independence, 2020

For a complete list of titles in this series, visit the
AMS Bookstore at **www.ams.org/bookstore/memoseries/**.